Hands-On Automated Machine Learning

自动机器学习入门与实践：使用Python

[美] Sibanjan Das　Umit Mert Cakmak　著
谢琼娟 译　　马勇 审校

华中科技大学出版社
中国·武汉

内容简介

AutoML可以将部分机器学习过程自动化，减轻数据科学从业者的工作负担，深受高级分析人员的喜爱。本书介绍搭建AutoML模块的基础知识，并通过练习帮助读者消化这些知识。读者将学习使用机器学习流水线自动实现数据预处理、特征选择、模型训练、模型优化等任务，学习应用auto-sklearn和MLBox等已有的自动化库，并且创建和扩展自定义的AutoML环节。阅读本书，你将对AutoML有更清晰的认识，能利用真实数据集完成自动化任务。书中知识可运用到实际的机器学习项目中，或者在机器学习竞赛中助你一臂之力。

图书在版编目(CIP)数据

自动机器学习入门与实践：使用Python / (美) 西班扬·达斯, (美) 乌米特·卡卡马克著；谢琼娟译. ——武汉：华中科技大学出版社, 2019.12

ISBN 978-7-5680-4952-8

Ⅰ.①自… Ⅱ.①西… ②乌… ③谢… Ⅲ.①软件工具－程序设计 Ⅳ.①TP311.561

中国版本图书馆CIP数据核字(2019)第262461号

Copyright© Packt Publishing 2018. First published in the English Language under the title Hands-On Automated Machine Learning.

湖北省版权局著作权合同登记 图字：17-2019-267号

书　　名	自动机器学习入门与实践：使用Python	
	Zidong Jiqi Xuexi Rumen yu Shijian: Shiyong Python	
作　　者	[美] Sibanjan Das　Umit Mert Cakmak	
译　　者	谢琼娟	
审　　校	马　勇	
策划编辑	徐定翔	
责任编辑	陈元玉	
责任监印	徐　露	

出版发行　华中科技大学出版社（中国·武汉）

　　　　　武汉市东湖新技术开发区华工科技园（邮编430223 电话027-81321913）

录　　排　武汉东橙品牌策划设计有限公司

印　　刷　湖北新华印务有限公司

开　　本　787mm × 960mm 1/16

印　　张　15.5

字　　数　383千字

版　　次　2019年12月第1版第1次印刷

定　　价　72.90元

本书若有印装质量问题，请向出版社营销中心调换

全国免费服务热线400-6679-118竭诚为您服务

版权所有 侵权必究

前言
Preface

亲爱的读者，欢迎来到自动**机器学习**（machine learning，ML）的世界。**自动机器学习**（automated ML，AutoML）的使命是将部分机器学习过程自动化。现成的 AutoML 工具可减轻数据科学从业者的工作负担，在高级分析领域中接受度很高。本书介绍搭建 AutoML 模块的基础知识，并通过实践促进读者快速吸收这些知识。

读者将学习使用机器学习流水线自动实现数据预处理、特征选择、模型训练、模型优化等任务。书中会讲解如何应用 auto-sklearn 和 MLBox 等已有自动化库，以及创建和扩展自定义 AutoML 学习组件。

阅读本书，你会对 AutoML 的各个方面有更清晰的认识，能利用真实数据集完成自动化任务。书中的知识可运用到机器学习项目实践中，或在机器学习竞赛中助你一臂之力。我们希望购买本书的读者都觉得物有所值，干货十足。

目标读者
Who This Book is For

本书尤其适合机器学习初学者（包括新晋数据科学家、数据分析师、机器学习爱好者）学习，同时也适合对搭建高速机器学习流水线感兴趣的机器学习工程师和数据专业人员阅读。

大纲
What This Book Covers

第 1 章　AutoML 简介。为理解 AutoML 打基础，介绍各种自动化学习库。

第 2 章　Python 机器学习简介。介绍机器学习概念，便于理解 AutoML 方法。

第 3 章　数据预处理。深入诠释各种数据预处理方法、自动化对象、如何自动化，也会介绍特征工具和 sklearn 预处理方法。

第 4 章　自动化算法选择。指出哪些算法适用于哪类数据集。介绍不同算法的计算难度和可扩展性，也会接触到一些依据训练和推理时间来确定使用哪种算法的方法。本章会演示 auto-sklearn，以及如何扩展引入新算法。

第 5 章　超参数优化。讲解自动化超参数优化的基础知识。

第 6 章　创建 AutoML 流水线。阐述如何将不同组件组合起来构建一个端到端的 AutoML 流水线。

第 7 章　深度学习探究。介绍诸多深度学习概念及其对 AutoML 的贡献。

第 8 章　机器学习和数据科学项目的重点。总结全文，并分享一些从多方面权衡 AutoML 复杂性和成本的信息。

充分利用本书
To Get the Most Out of This Book

阅读本书唯一需要准备的是对机器学习的求知欲。除此之外，如果你以前接触过 Python 编程和机器学习基础知识，则能更好地利用本书，但这并非必备前提。学习本书，请提前安装 Python 3.5 和 Jupyter Notebook。

若具体章节中有特别要求，则会在该章第一节中提出。

下载代码示例
Download the Example Code Files

你可通过账号登录 www.packtpub.com，下载本书的代码示例。如果你在其他渠道购买本书，可访问 www.packtpub.com/support 进行注册，通过邮件接收代码示例。

下载代码示例的步骤如下。

（1）在 www.packtpub.com 登录或注册。

（2）选择 Support 页签。

（3）点击 Code Downloads & Errata。

（4）在搜索框中输入本书名称，并根据提示进行操作。

下载代码示例后，使用以下工具最新版进行解压或提取。

- Windows 系统：WinRAR/7-Zip。
- Mac 系统：Zipeg/iZip/UnRarX。
- Linux 系统：7-Zip/PeaZip。

本书涉及的代码包已上传到 GitHub，地址：https://github.com/PacktPublishing/Hands-On-Automated-Machine-Learning。如果代码有更新，则会在当前 GitHub 仓库中进行更新。

在 https://github.com/PacktPublishing/ 还可找到丰富的其他书目和视频中用到的代码包。去看看吧！

下载彩色图片
Download the Color Images

本书提供 PDF 版，包含本书中用到的彩色版截图或图表，下载链接为：https://www.packtpub.com/sites/default/files/downloads/HandsOnAutomatedMachineLearning_ColorImages.pdf。

文本约定
Conventions Used

本书中用到了一些文本约定。

CodeInText（文本代码）：表示代码文本、数据库表格名称、文件夹名称、文件名、文件扩展名、路径、虚拟 URL、用户输入及 Twitter 账号。比如："使用 sklearn.preprocessing 模块中的 StandardScaler 对 statisfaction_level 一栏的值进行标准化处理。"

代码片段如下：

```
{'algorithm': 'auto',
 'copy_x': True,
 'init': 'k-means++',
 'max_iter': 300,
 'n_clusters': 2,
 'n_init': 10,
 'n_jobs': 1,
 'precompute_distances': 'auto',
 'random_state': None,
 'tol': 0.0001,
 'verbose': 0}
```

命令行输入或输出如下：

```
pip install nltk
```

黑体：表示新术语、重要词汇，或在屏幕上看到的词语。比如，菜单或对话框中的内容会显示为**黑体**。例如："出现一个 **NLTK Downloader** 下载弹窗。在 **Identifier** 中选择**全部**，等待安装完成。"

警告或重要的注释，用此图标表示。

提示或技巧，用此图标表示。

目录

Table of Contents

第 1 章 AutoML 简介 1
- 1.1 机器学习的范围 2
- 1.2 什么是 AutoML 4
- 1.3 为什么和怎么用 AutoML 10
- 1.4 何时需要将机器学习自动化 11
- 1.5 能学到什么 11
  - AutoML 的核心环节 11
  - 为每个环节构建原型子系统 13
  - 组合形成端到端的 AutoML 系统 13
- 1.6 AutoML 库概述 13
  - Featuretools 13
  - auto-sklearn 16
  - MLBox 18
  - TPOT 21
- 1.7 总结 23

第 2 章 Python 机器学习简介 25
- 2.1 技术要求 26
- 2.2 机器学习 26
  - 机器学习流程 27
  - 监督学习 27
  - 无监督学习 28
- 2.3 线性回归 28
  - 什么是线性回归 28

　　　　在哪里用线性回归 31
　　　　采用什么方法实现线性回归 31
　2.4　重要评估指标——回归算法 37
　2.5　逻辑回归 39
　　　　什么是逻辑回归 39
　　　　在哪里使用逻辑回归 39
　　　　使用什么方法实现逻辑回归 39
　2.6　重要评估指标——分类算法 44
　2.7　决策树 46
　　　　什么是决策树 47
　　　　在哪里使用决策树 47
　　　　使用什么方法实现决策树 47
　2.8　支持向量机 49
　　　　什么是 SVM 50
　　　　在哪里使用 SVM 50
　　　　使用什么方法实现 SVM 50
　2.9　K 近邻算法 52
　　　　什么是 KNN 52
　　　　在哪里使用 KNN 52
　　　　使用什么方法实现 KNN 52
　2.10　集成方法 54
　　　　集成模型是什么 54
　2.11　分类器结果对比 59
　2.12　交叉验证 60
　2.13　聚类 61
　　　　什么是聚类 61
　　　　在哪里使用聚类 62
　　　　用什么方法实现聚类 62
　　　　层次聚类 63
　　　　划分聚类（KMeans） 64
　2.14　总结 66

第 3 章　数据预处理 67
　3.1　技术要求 68
　3.2　数据转换 68

		数值型数据的转换	68
		类别型数据转换	88
		文本预处理	93
	3.3	特征选择	97
		低方差特征排除	97
		单变量特征选择	99
		递归特征消除	99
		随机森林特征选择	100
		降维特征选择	101
	3.4	特征生成	103
	3.5	总结	105
第4章	自动化算法选择		107
	4.1	技术要求	108
	4.2	计算复杂度	108
		大 O 表示法	108
	4.3	训练时间和推理时间的区别	110
		训练时间和推理时间的简化度量	111
		Python 代码分析	113
		性能统计数据可视化	114
		从头开始实现 KNN	116
		逐行分析 Python 脚本	117
	4.4	线性与非线性	119
		画出决策边界	119
		逻辑回归的决策边界	120
		随机森林的决策边界	122
		常用机器学习算法	123
	4.5	必要特征转换	124
	4.6	监督机器学习	125
		auto-sklearn 默认配置	126
		找出产品线预测的最佳机器学习流水线	127
		找出网络异常检测的最佳机器学习流水线	131
	4.7	无监督 AutoML	132
		常用聚类算法	132
		使用 sklearn 创建样本数据集	133

　　　　　k-means 算法实践 137
　　　　　DBSCAN 算法实践 141
　　　　　凝聚聚类算法实践 143
　　　　　无监督学习的简单自动化 144
　　　　　高维数据集视觉化 147
　　　　　主成分分析实践 149
　　　　　t-SNE 实践 152
　　　　　简单成分叠加以改善流水线 155
　　4.8　总结 157

第 5 章　超参数优化 159
　　5.1　技术要求 160
　　5.2　超参数 161
　　5.3　热启动 173
　　5.4　贝叶斯超参数优化 174
　　5.5　示例系统 175
　　5.6　总结 178

第 6 章　创建 AutoML 流水线 179
　　6.1　技术要求 180
　　6.2　机器学习流水线简介 180
　　6.3　简单的流水线 182
　　6.4　函数转换器 184
　　6.5　复杂流水线 187
　　6.6　总结 190

第 7 章　深度学习探究 191
　　7.1　技术要求 192
　　7.2　神经网络概览 192
　　　　　神经元 194
　　　　　激活函数 194
　　7.3　使用 Keras 的前馈神经网络 198
　　7.4　自编码器 201
　　7.5　卷积神经网络 205
　　　　　为什么使用 CNN 206
　　　　　什么是卷积 206

	什么是过滤器	206
	卷积层	207
	ReLU 层	207
	池化层	207
	全连接层	208
7.6	总结	210

第 8 章 机器学习和数据科学项目的重点 211
- 8.1 机器学习搜索 211
- 8.2 机器学习的权衡 221
- 8.3 典型数据科学项目的参与模型 222
- 8.4 参与模型的阶段 223
  - 业务理解 224
  - 数据理解 225
  - 数据准备 226
  - 建模 226
  - 评估 227
  - 部署 228
- 8.5 总结 228

作者简介 230

索引 231

第 1 章

AutoML 简介
Introduction to AutoML

过去十年，科学和技术领域发生了翻天覆地的变化。2007 年，第一部 iPhone 面世，在那之前，手机都用物理键盘。触摸屏并不是最新的发明，因为 Apple 公司已经有过类似的原型产品，而且 IBM 公司早在 1994 年就推出了最早的智能手机 Simon 个人通信器（Simon personal communicator）。Apple 公司的想法是制造一部集听音乐、看视频、浏览网站和 GPS 导航等全部多媒体功能于一体的设备，第一代 iPhone 的计算能力已经足以支撑以上这些功能。技术发展到今天的程度，速度惊人。第一代 iPhone 诞生十年后，现在的 iPhone 除了基本功能，已经可以识别人脸，可以认出动物、车辆和食品等实物，还能理解自然语言，跟人对话。

现在又有了可打印器官的 3D 打印机、自动驾驶汽车、可编组飞行的无人机、基因编辑、可重复利用的火箭和会后空翻的机器人。这些都不再是科幻小说里读到的情景了，而是实实在在发生在现实生活中的例子。过去的想象，已变成如今的现实。大家甚至开始谈论人工智能（artificial intelligence，AI）对人类的威胁。许多前沿科学家，如斯蒂芬·霍金，都在警告人类可能被 AI 之类的生命体毁灭。

AI 和机器学习（machine learning，ML）在过去几年里成了最热门的话题。过去十多年里，机器学习算法取得诸多成果和重大进步，如 Google 公司的 AlphaGo 击败了全球第一的人类围棋手柯杰。这并不是机器学习算法第一次击败人类。有些精细场景，比如识别不同动物的物种，机器算法常常比人类更厉害。

这种显著进步引起了商业界的浓厚兴趣。这些技术虽然听起来是独立的学术研究，但其实有着巨大的商业意义。各行业的企业都想要利用这些算法，努力适应不断变革的技术场景。大家都意识到，谁能把这些技术运用到业务中，谁就能拔得头筹。本章讲解什么是机器学习和 AutoML，内容包括以下几个方面。

- 机器学习的范围。
- 什么是 AutoML。
- 为什么和怎么用 AutoML。
- 何时需要 AutoML。
- AutoML 库概览。

1.1 机器学习的范围
Scope of Machine Learning

机器学习和预测分析有助于企业关注重点领域、提前预知问题、节约成本、提高收入。这是继**商业智能**（business intelligence，BI）之后的自然演变。商业智能多采用仪表盘，展示各种**关键表现指标**（key performance indicators，KPI）和性能指标，以便系统地监控业务流程，支持企业做出更好的决策。

商业智能工具可深入挖掘组织中的历史数据，发现趋势，理解季节效应，调查非常规事件，等等。这类工具也提供实时分析，支持设置警告和提示，以便精准地管理事件。然而，商业智能工具已经无法满足现代商业的需求。为什么？商业智能工具主要处理历史数据和近实时数据，但它解答不了未来的问题，比如它无法回答如下问题。

- 生产线中的哪台机器可能出故障？
- 哪个客户可能投奔竞争对手？
- 哪个公司的股价明天会上涨？

解答企业现在想知道的这类问题，机器学习和预测分析就能派上用场了。

不过要小心！机器学习模型并不一定能保证获得准确的结果。尽管机器学习的发展速度超乎想象，但我们还是需要找到正确的应用方向和领域，才能让机器学习发挥真正的作用，从而创造实用价值。

要发挥机器学习的作用，最好先从具备以下条件的小项目入手。

- 有相对容易的决策流程。
- 你熟悉各种假设条件。
- 你熟悉已有的数据。

这里的关键是项目范围和执行步骤要有明确的定义。不过，从小项目做起，不代表愿景也小。要始终考虑未来的可扩展性，支持慢慢加码到大数据源。

机器学习算法有很多种，每一种都为解决特定问题而设计，各有利弊。机器学习领域的研究一直在发展，从业者每天都在提出新方法，不断扩张新边界。结果是，研究人员很容易迷失在爆炸的信息中，尤其在开发机器学习应用的时候，建模的每个阶段都有许多可选的工具和技术。为了更容易构建机器学习模型，需要将整个过程分解成多个小阶段。**自动机器学习**（automated ML，AutoML）流水线中有许多动态的部分，如特征预处理、特征选择、模型选择和超参数优化。其中每个部分都要悉心处理，才能成功交付项目。

稍后我们会详细介绍机器学习的概念，现在先讲讲为什么要关注 AutoML。

解决问题的工具和技术越来越多，有时反而成了困扰，因为调研和探索某个问题的解决方法会耗费大量时间。机器学习的问题也一样。高性能机器学习模型的搭建包含若干精巧的小步骤，步步相连，构成机器学习流水线，然后合理泛化到生产环境中。

流水线中涉及许多步骤，因而可能变得冗长繁杂。每一步都有许多方法可选，而且这些方法还能排列组合，所以必须系统性地验证机器学习流水线中的环节。

这时候就该 AutoML 出场了！

1.2 什么是 AutoML
What is AutoML?

自动机器学习（AutoML）将特征预处理、模型选择和超参数优化等常用步骤自动化，以简化机器学习的建模流程。接下来的章节会详细介绍这些步骤，并且会教读者动手构建一套 AutoML 系统，从而对 AutoML 工具和库有更深刻的理解。

在开始之前，有必要回顾一下什么是机器学习模型，以及如何训练模型。

机器学习算法对数据进行处理，识别特定的模式，这一学习过程称为**模型训练**（model training）。模型训练的结果是机器学习模型。有了机器学习模型，你不用制定明确的规则，它就可针对数据提出见解或解答。

在实际应用机器学习模型时，需要输入大量数据，用于算法训练。训练后的成果是可用于预测的机器学习模型。这种预测可根据服务器当前状态来确定它未来四个小时是否需要维护，或者判断客户会不会投向竞争对手。

有时待解决的问题本身都没有明确定义，甚至我们都不知道需要什么样的答案。在这种情况下，机器学习模型可帮助探索数据集，比如识别行为相似的客户群，或者根据不同股票之间的关联关系发现股票的层级结构。

模型划分出客户群后，有什么用？至少可以知道：同一群体的客户有哪些相似的特征，比如年龄、职业、婚姻状况、性别、喜好、日常消费习惯、总消费额等。不同群体的客户是彼此不同的。有了这些信息，我们就可以针对每个群体推送不同的广告。

可以使用简单的数学术语说明这一流程。设有数据集 X，包含 n 个样本。样本可代表客户或不同的动物。通常，每个样本都是一个实数集，称为**特征**（feature），比如，一位 35 岁的女性客户在商店消费了 12000 美元，可以用向量（0.0, 35.0, 12000.0）表示。注意，这里性别是用 *0.0* 表示的，男性客户可以用 *1.0* 表示。向量的大小称为维度，通常用 m 表示。这是一个大小为 3 的向量，即三维数据集。

根据问题的类型不同，需要为每个样本添加标签。假如这是一个二分类问题，可以用 1.0 和 0.0 标识样本，那这个新变量就叫**标签**（label）或**目标**（target）变量。目标变量一般用 y 表示。

有了 x 和 y，机器学习模型就可以看成是一个权重为 w（模型参数）的函数 f：

$$f(x;w)$$

模型参数是在训练过程中学到的，还有些参数是在训练开始前设置好的，这种参数称为**超参数**（hyperparameter），稍后会对这个概念进行解释。

数据集的特征应该先进行预处理，再用于模型训练。比如，有些机器学习模型隐含的假设是特征呈正态分布。在许多真实场景中，特征并不是正态分布的，那么就需要借助一些特征转换法，如对数转换，把特征调整为正态分布。

特征处理完成，且模型超参数也设置好后，就开始模型训练了。模型训练结束后，会学到模型参数，可以用来预测新数据的目标变量。模型做出的预测通常表示为 $\hat{y}$：

$$\hat{y} = f(x;w)$$

训练过程中发生了什么？由于我们已知训练数据集的标签，可以将当前模型预测出的结果与原标签做对比，据此反复调整模型参数。

这种对比是基于**损失函数**（也称代价函数，loss function 或 cost function）进行的，表示为 $L(\hat{y}, y)$。损失函数可表示预测的失真度。常见的损失函数有平方损失（square loss）、合页损失（hinge loss）、逻辑损失（logistic loss）和交叉熵损失（cross-entropy loss）。

模型训练完成后，可以用 test 测试数据测试机器学习模型的性能。test 数据是训练过程中没有用过的数据集，用于检验模型的泛化能力。评估模型的表现可以用不同的指标；根据得到的结果再重复前面的步骤，反复调整，得到更好的性能。

现在，你应该对训练机器学习模型的工作原理有了大概的认识。

那么，AutoML 是什么呢？说起 AutoML，大多数情况下是指自动化数据准备（即特征预处理、特征提取、特征选择）和模型训练（模型选择、超参数优化）。这个过程中的每一步有多少可选项，根据问题类型不同而有很大差别。

AutoML 能让研究者和实践者依据每一步的选项自动化构建出机器学习流水线，从而找到高性能机器学习模型。

图 1-1 展示了一个典型的机器学习模型的生命周期，每个步骤均有几个示例。

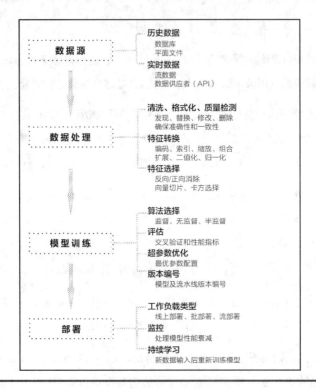

图 1-1　机器学习模型生命周期

数据源多种多样，如平面文件、数据库、API。采集数据后，应该将数据处理好，供机器学习使用，通用的操作步骤有数据清洗、格式化、特征转换、特征选择等。数据处理过后，生成的最终数据集就可用于机器学习，并依据数据集选择使用什么算法。选出的算法可通过交叉验证和超参数优化等进行验证和优化。最终的模型可选择线上部署、批部署或流部署。模型开始实际使用后，要监控其表现，如有需要，可采取必要的措施，如重新训练、重新评估和重新部署。

构建机器学习模型时，首先要对目标领域进行调研，明确目标。此过程中涉及许多步骤，需在实际开始工作前计划好并进行文档记录。关于完整项目管理过程的更多信息，可参考 CRISP-DM 模型（https://en.wikipedia.org/wiki/Cross-industry_standard_process_for_data_mining），项目管理是模型成功应用的重要因素，但是不在本书讨论范围内。

构建机器学习流水线，通常会遇到多种数据源，如关系型数据库或非结构化数据文件，可从中获取历史数据，也可从各处将流数据引入系统中。

调研哪些数据源对某个任务有价值，继续进入数据处理阶段，其中包括大量的清洗、格式化和数据质量检测，然后是特征的转换和选择。

数据集准备就绪，还需要考虑用哪一个或多个机器学习模型合适。训练多个模型，评估模型，找出最佳的超参数设置。整个过程中不要忘记版本编号，以便记录所有变更。实验结束后，可生成一个高性能机器学习流水线，每一步的性能都经过了优化。用性能最好的机器学习流水线在生产环境进行试运行，即进入部署阶段实施流水线。

实施机器学习流水线意味着要选择一种部署类型。有些负载是用于对数据集中的数据进行批处理的，则需要批部署。还有些负载可能用于处理各种数据源提供的实时数据，则需要流部署。

仔细检查每一步，会发现有许多数据处理选项和训练步骤。首先，选择合适的方法和算法，然后对所选方法和算法进行超参数优化，针对特定问题实现最好的性能。

举一个简单的例子，假设已完成模型训练这一步，接下来要选择一套模型进行实验。为了简便起见，假设我们只选用一种称为 k-means 的实验算法，那么主要的任务就是调参数。

k-means 算法可将相似的数据点聚类。以下代码使用了 scikit-learn 库，可以使用 pip 安装（https://scikit-learn.org/stable/install.html）。无法完全看懂也没关系。

```
# Sklearn has convenient modules to create sample data.
# make_blobs will help us to create a sample data set suitable for
clustering
from sklearn.datasets.samples_generator import make_blobs
```

```
X, y = make_blobs(n_samples=100, centers=2, cluster_std=0.30,
random_state=0)

# Let's visualize what we have first
import matplotlib.pyplot as plt
import seaborn as sns

plt.scatter(X[:, 0], X[:, 1], s=50)
```

上述代码输出结果如图 1-2 所示。

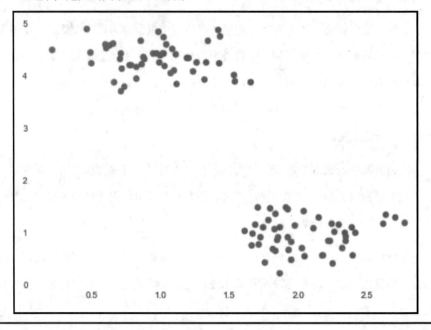

图 1-2 聚类

可以看到，图 1-2 中有两个聚类：

```
# We will import KMeans model from clustering model family of Sklearn
from sklearn.cluster import KMeans

k_means = KMeans(n_clusters=2)
k_means.fit(X)
predictions = k_means.predict(X)

# Let's plot the predictions
plt.scatter(X[:, 0], X[:, 1], c=predictions, cmap='brg')
```

上述代码输出结果如图 1-3 所示。

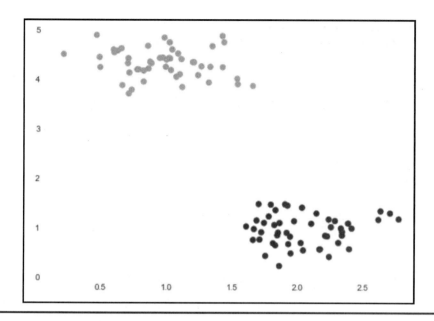

图 1-3　k-means 聚类分布

很好！算法如愿生效。敏锐的读者们可能已经发现 k-means 模型有一个称为 n_clusters 的参数。把值 2 赋给这个参数后，算法就会尝试将数据集分成两个聚类。你也许已经猜到，本例中 k-means 的超参数就是聚类数量。k-means 模型需要在训练前就知道这个参数。

不同的算法有不同的超参数，如决策树的超参数是树深，神经网络的超参数是隐层数和学习率，Lasso 的超参数是 α（alpha），或**支持向量机**（support vector machines，SVM）的超参数是 C、kernel 和 γ（gamma）。

可以使用 get_params 方法查看 k-means 模型有多少个参数：

```
k_means.get_params ()
```

输出的是全部可优化的参数列表：

```
{'algorithm': 'auto',
 'copy_x': True,
 'init': 'k-means++',
 'max_iter': 300,
 'n_clusters': 2,
 'n_init': 10,
 'n_jobs': 1,
```

```
'precompute_distances': 'auto',
'random_state': None,
'tol': 0.0001,
'verbose': 0}
```

大多数实际用例中，我们既没有资源，也没有时间尝试所有的步骤组合。

这个时候AutoML库就可登场了，它会准备好各种机器学习流水线实验，覆盖从数据采集、数据处理、建模到推理的全部步骤。

1.3 为什么和怎么用 AutoML
Why Use AutoML and How Does It Help?

互联网上有许多关于机器学习的教程，通常样本数据集都是干净的，格式标准，直接可用于算法，因为大部分教程的目标主要是展示某些特定的工具、库或SaaS（Software as a Service）服务的能力。

而现实数据集的类型和大小各异。2017 年，Kaggle 基于收集到的 16000 份反馈完成了一份数据科学和机器学习现状（The State of Data Science and Machine Learning）的行业调查，其中提到最常用到的三种数据类型是关系型数据、文本数据和图像数据。

此外，Kaggle 的调研还指出，脏数据是最棘手的问题之一。数据集不干净的话，需要耗费大量的时间精力进行特殊处理，之后才能用于机器学习模型，包括数据清洗、修正，还要人工修补特征数据。数据处理在任何数据科学项目中都是最耗时的环节。

那高性能机器学习模型选择，以及模型训练、模型验证和测试阶段的超参数优化选择呢？都是特别重要的步骤，同样有很多方法可选。

把恰当的部件组合起来进行数据集的处理和建模，就形成机器学习流水线，待实验的流水线会越来越多。

每一步都结合时间和软件/硬件资源的限制，系统性考虑所有选项，是成功构建高性能机器学习流水线所不可或缺的。

在为解决某个问题而自动构建机器学习流水线的过程中，AutoML 系统帮助我们确定哪些是最可靠的方法，并高效执行这些方法，从而选出最优模型。

1.4 何时需要将机器学习自动化
When Do You Automate ML

能上手构建机器学习流水线之后，你就会发现仍有许多准备特征和调参的必要流程。你可能更熟悉某些方法，更清楚哪些技术在不同的参数配置下配合得更好。

你会在不同的项目中通过做各种实验评估处理和流水线建模而积累更多的经验，逐步迭代地优化整个流水线。但如果开始没有头绪，整个流程就会很快扭曲错乱。

当处理众多动态环节和大量参数有困难时，AutoML 的必要性就体现出来了。AutoML 能有序地帮你突出重围，更专注于设计和实现细节。

1.5 能学到什么
What Will You Learn?

本书中，你会从理论和实践两个方面学习 AutoML 系统。更重要的是，你可从头开发一套 AutoML 系统，练习实用的技能。

AutoML 的核心环节
Core Components of AutoML Systems

本节会回顾 AutoML 的以下核心环节。

- 自动化特征预处理。
- 自动化算法选择。
- 超参数优化。

更好地了解核心环节，有助于在头脑中形成一套 AutoML 系统的全局图。

自动化特征预处理

处理机器学习问题时，通常有一个关系型数据集，其中包含各种类型的数据，而且应该在训练机器学习算法之前把各种数据都处理好。

比如，处理数值型数据，可用最小-最大缩放法或方差缩放法对数据进行缩放。

处理文本型数据，可去除停止词，如 *a*、*an* 和 *the* 等[1]，进行词干提取、解析、标记替换等处理。

对于类别型数据，需用独热编码、伪编码和特征哈希等方法进行转换。

那如果特征非常多呢？比如，有几千个特征，其中多少是真正有用的？用**主成分分析法**（principal component analysis，PCA）等方法进行降维是不是更好呢？

如果有视频、音频和图像等不同格式的数据呢？怎么处理这些数据？

例如，图像数据可进行转换，如将图像调整成常见形状，也可切割成不同的分区。

特征预处理方法非常多，适当转换后，机器学习算法效果更好。武器库中如果有一套灵活的 AutoML 系统，就可以更智能地进行不同组合的实验，在项目中节约许多珍贵的时间和金钱。

自动化算法选择

特征处理完成后，要找出一套合适的算法进行训练和评估。

每个机器学习算法都能解决一些问题。先考虑聚类算法，如 k-means、层次聚类、谱聚类和 DBSCAN[2]。大家很熟悉 k-means，但其他算法呢？每种算法都有应用领域，由于数据集的分布特性不同，某种算法在某一领域可能优于其他算法。

AutoML 流水线有助于从一系列适用算法中选出最合适的算法，解决某个特定问题。

---

[1] 中文则是"的"、"地"、"得"、"了"等非实义词。
[2] 具有噪声的基于密度的聚类方法。

超参数优化

每一种机器学习算法都有一个或多个超参数,如你已经熟悉的 k-means。不只机器学习算法有超参数,特征处理方法也有超参数,也需要优化。

超参数优化对模型的成功是至关重要的,AutoML 流水线可帮助定义一系列用于实验的超参数,以找出最好的机器学习流水线。

为每个环节构建原型子系统
Building Prototype Subsystems for Each Component

阅读本书的过程,也是你从零开始构建 AutoML 系统的每一个核心环节并观察各个环节之间如何互动的过程。

从头构建这些系统,可加深对整个流程的理解,也可促进对常用 AutoML 库内部工作机制的了解。

组合形成端到端的 AutoML 系统
Putting It all Together As an End-to-End AutoML System

通读全书后,你会对核心环节及这些环节如何配合构建出机器学习流水线有更充分的认识。然后,运用自己的知识,从头开始写 AutoML 流水线,随时调整,更好地解决问题。

1.6　AutoML 库概述
Overview of AutoML Libraries

常用的 AutoML 库很多,本章节中会简要描述数据科学社区中常用的库。

Featuretools
Featuretools

Featuretools(https://www.featuretools.com/)是一个很好的自动化特征工程库,适用于关系型数据和事务型数据。库中引进了一个概念,称为**深度特征合成**

（deep feature synthesis，DFS）。如果有多个定义了关系的数据集，如依据唯一标识栏定义为父子关系，DFS 会进行一些如求和、计数、取平均、求众数、求标准差的计算，创建出新的特征。我们来看一个小例子，有两个表格，一个展示数据库信息，另一个展示每个数据库的数据库事务：

```python
import pandas as pd

# First dataset contains the basic information for databases.
databases_df = pd.DataFrame({"database_id": [2234, 1765, 8796, 2237, 3398],
"creation_date": ["2018-02-01", "2017-03-02", "2017-05-03", "2013-05-12",
"2012-05-09"]})

databases_df.head()
```

输出如图 1-4 所示。

	creation_date	database_id
0	2018-02-01	2234
1	2017-03-02	1765
2	2017-05-03	8796
3	2013-05-12	2237
4	2012-05-09	3398

图 1-4　数据库信息

以下是数据库事务的代码：

```python
# Second dataset contains the information of transaction for each database
id
db_transactions_df = pd.DataFrame({"transaction_id": [26482746, 19384752,
48571125, 78546789, 19998765, 26482646, 12484752, 42471125, 75346789,
16498765, 65487547, 23453847, 56756771, 45645667, 23423498, 12335268,
76435357, 34534711, 45656746, 12312987],
            "database_id": [2234, 1765, 2234, 2237, 1765, 8796, 2237,
8796, 3398, 2237, 3398, 2237, 2234, 8796, 1765, 2234, 2237, 1765, 8796,
2237],
            "transaction_size": [10, 20, 30, 50, 100, 40, 60, 60, 10,
20, 60, 50, 40, 40, 30, 90, 130, 40, 50, 30],
            "transaction_date": ["2018-02-02", "2018-03-02",
"2018-03-02", "2018-04-02", "2018-04-02", "2018-05-02", "2018-06-02",
"2018-06-02", "2018-07-02", "2018-07-02", "2018-01-03", "2018-02-03",
"2018-03-03", "2018-04-03", "2018-04-03", "2018-07-03", "2018-07-03",
"2018-07-03", "2018-08-03", "2018-08-03"]})

db_transactions_df.head()
```

输出如图 1-5 所示。

	database_id	transaction_date	transaction_id	transaction_size
0	2234	2018-02-02	26482746	10
1	1765	2018-03-02	19384752	20
2	2234	2018-03-02	48571125	30
3	2237	2018-04-02	78546789	50
4	1765	2018-04-02	19998765	100

图 1-5　数据库事务

实例的代码如下：

```
# Entities for each of datasets should be defined
entities = {
"databases" : (databases_df, "database_id"),
"transactions" : (db_transactions_df, "transaction_id")
}

# Relationships between tables should also be defined as below
relationships = [("databases", "database_id", "transactions",
"database_id")]

print(entities)
```

以上代码输出如下：

```
{'databases': (  creation_date  database_id  database_size
0    2018-02-01         2234            50
1    2017-03-02         1765           120
2    2017-05-03         8796           100
3    2013-05-12         2237            30
4    2012-05-09         3398            30, 'database_id'), 'transactions': (   database_id transaction_date  transaction_id  transaction_size
0          2234       2018-02-02        26482746                10
1          1765       2018-03-02        19384752                20
2          2234       2018-03-02        48571125                30
3          2237       2018-04-02        78546789                50
4          1765       2018-04-02        19998765               100
5          8796       2018-05-02        26482646                40
6          2237       2018-06-02        12484752                60
7          8796       2018-06-02        42471125                60
8          3398       2018-07-02        75346789                10
9          2237       2018-07-02        16498765                20
10         3398       2018-01-03        65487547                60
11         2237       2018-02-03        23453847                50
12         2234       2018-03-03        56756771                40
13         8796       2018-04-03        45645667                40
14         1765       2018-04-03        23423498                30
15         2234       2018-07-03        12335268                90
16         2237       2018-07-03        76435357               130
17         1765       2018-07-03        34534711                40
18         8796       2018-08-03        45656746                50
19         2237       2018-08-03        12312987                30, 'transaction_id')}
```

以下代码片段创建特征矩阵和特征定义：

```
# There are 2 entities called 'databases' and 'transactions'

# All the pieces that are necessary to engineer features are in place, you
can create your feature matrix as below

import featuretools as ft

feature_matrix_db_transactions, feature_defs = ft.dfs(entities=entities,
relationships=relationships,
target_entity="databases")
```

输出生成特征，如图 1-6 所示。

database_id	database_size	SUM(transactions.transaction_size)	STD(transactions.transaction_size)	MAX(transactions.tr
1765	120	190	31.124749	
2234	50	170	29.474565	
2237	30	340	35.433819	
3398	30	70	25.000000	
8796	100	190	8.291562	

图 1-6　用数据库和事务实例生成特征

查看 features_defs，可看到全部的特征定义：

feature_defs

输出如下：

```
[<Feature: database_size>,
 <Feature: SUM(transactions.transaction_size)>,
 <Feature: STD(transactions.transaction_size)>,
 <Feature: MAX(transactions.transaction_size)>,
 <Feature: SKEW(transactions.transaction_size)>,
 <Feature: MIN(transactions.transaction_size)>,
 <Feature: MEAN(transactions.transaction_size)>,
 <Feature: COUNT(transactions)>,
 <Feature: DAY(creation_date)>,
 <Feature: YEAR(creation_date)>,
 <Feature: MONTH(creation_date)>,
 <Feature: WEEKDAY(creation_date)>,
 <Feature: NUM_UNIQUE(transactions.DAY(transaction_date))>,
 <Feature: NUM_UNIQUE(transactions.YEAR(transaction_date))>,
 <Feature: NUM_UNIQUE(transactions.MONTH(transaction_date))>,
 <Feature: NUM_UNIQUE(transactions.WEEKDAY(transaction_date))>,
 <Feature: MODE(transactions.DAY(transaction_date))>,
 <Feature: MODE(transactions.YEAR(transaction_date))>,
 <Feature: MODE(transactions.MONTH(transaction_date))>,
 <Feature: MODE(transactions.WEEKDAY(transaction_date))>]
```

这就展示了如何轻松地基于关系型和事务型数据集生成特征。

auto-sklearn

auto-sklearn

scikit-learn 有一套很好的开发机器学习模型和流水线的 API。scikit-learn 的 API 非常一致和成熟；习惯使用后，auto-sklearn（http://automl.github.io/auto-sklearn/stable/）也会一样好用，因为这就是一套简易版的 scikit-learn 估算器替代工具。

下面看一个小例子：

```
# Necessary imports
import autosklearn.classification
import sklearn.model_selection
import sklearn.datasets
import sklearn.metrics
from sklearn.model_selection import train_test_split
# Digits dataset is one of the most popular datasets in machine learning
community.
# Every example in this datasets represents a 8x8 image of a digit.
X, y = sklearn.datasets.load_digits(return_X_y=True)

# Let's see the first image. Image is reshaped to 8x8, otherwise it's a
vector of size 64.
X[0].reshape(8,8)
```

输出如下：

```
array([[ 0.,  0.,  5., 13.,  9.,  1.,  0.,  0.],
       [ 0.,  0., 13., 15., 10., 15.,  5.,  0.],
       [ 0.,  3., 15.,  2.,  0., 11.,  8.,  0.],
       [ 0.,  4., 12.,  0.,  0.,  8.,  8.,  0.],
       [ 0.,  5.,  8.,  0.,  0.,  9.,  8.,  0.],
       [ 0.,  4., 11.,  0.,  1., 12.,  7.,  0.],
       [ 0.,  2., 14.,  5., 10., 12.,  0.,  0.],
       [ 0.,  0.,  6., 13., 10.,  0.,  0.,  0.]])
```

绘制几个图像，看看效果：

```
import matplotlib.pyplot as plt

number_of_images = 10
images_and_labels = list(zip(X, y))

for i, (image, label) in enumerate(images_and_labels[:number_of_images]):
    plt.subplot(2, number_of_images, i + 1)
    plt.axis('off')
    plt.imshow(image.reshape(8,8), cmap=plt.cm.gray_r, interpolation='nearest')
    plt.title('%i' % label)

plt.show()
```

运行上述代码片段，输出如下：

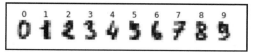

拆分数据集，对数据进行训练和测试：

```
# We split our dataset to train and test data
X_train, X_test, y_train, y_test = train_test_split(X, y, random_state=1)

# Similarly to creating an estimator in Scikit-learn, we create
AutoSklearnClassifier
```

```python
automl = autosklearn.classification.AutoSklearnClassifier()

# All you need to do is to invoke fit method to start experiment with
# different feature engineering methods and machine learning models
automl.fit(X_train, y_train)

# Generating predictions is same as Scikit-learn, you need to invoke
predict method.
y_hat = automl.predict(X_test)

print("Accuracy score", sklearn.metrics.accuracy_score(y_test, y_hat))
# Accuracy score 0.98
```

很简单，对不对？

MLBox
MLBox

MLBox（http://mlbox.readthedocs.io/en/latest/）也是一个 AutoML 库，支持分布式数据处理、清洗、格式化，以及如 LightGBM 和 XGBoost 等的最新算法。此库也支持模型叠加，可以将模型的信息集集成为一个新模型，拥有比单个模型更好的性能。

这里有应用示例，如下：

```
# Necessary Imports
from mlbox.preprocessing import *
from mlbox.optimisation import *
from mlbox.prediction import *
import wget
file_link =
'https://apsportal.ibm.com/exchange-api/v1/entries/8044492073eb964f46597b4b
e06ff5ea/data?accessKey=9561295fa407698694b1e254d0099600'
file_name = wget.download(file_link)

print(file_name)
# GoSales_Tx_NaiveBayes.csv
```

GoSales 数据集中包含顾客及其产品偏好的信息：

```
import pandas as pd
df = pd.read_csv('GoSales_Tx_NaiveBayes.csv')
df.head()
```

执行以上代码，输出如图 1-7 所示的 GoSales 数据集。

	PRODUCT_LINE	GENDER	AGE	MARITAL_STATUS	PROFESSION
0	Personal Accessories	M	27	Single	Professional
1	Personal Accessories	F	39	Married	Other
2	Mountaineering Equipment	F	39	Married	Other
3	Personal Accessories	F	56	Unspecified	Hospitality
4	Golf Equipment	M	45	Married	Retired

图 1-7　GoSales 数据集

从 GoSales 数据集中去掉一个 target 列，创建一个 test 集：

```
test_df = df.drop(['PRODUCT_LINE'], axis = 1)

# First 300 records saved as test dataset
test_df[:300].to_csv('test_data.csv')

paths = ["GoSales_Tx_NaiveBayes.csv", "test_data.csv"]
target_name = "PRODUCT_LINE"

rd = Reader(sep = ',')
df = rd.train_test_split(paths, target_name)
```

输出类似以下内容：

```
reading csv : GoSales_Tx_NaiveBayes.csv ...
cleaning data ...
CPU time: 0.48662495613098145 seconds

reading csv : test_data.csv ...
cleaning data ...
CPU time: 0.43369483947753906 seconds

> Number of common features : 4

gathering and crunching for train and test datasets ...
reindexing for train and test datasets ...
dropping training duplicates ...
dropping constant variables on training set ...

> Number of categorical features: 3
> Number of numerical features: 1
> Number of training samples : 2842
> Number of test samples : 300

> You have no missing values on train set...

> Task : classification
Personal Accessories        762
Camping Equipment           662
Mountaineering Equipment    550
Golf Equipment              440
Outdoor Protection          428
Name: PRODUCT_LINE, dtype: int64

encoding target ...
```

Drift_thresholder 会将 train 和 test 数据集之间的 ID 和漂移变量去掉:

```
dft = Drift_thresholder()
df = dft.fit_transform(df)
```

输出如下:

```
computing drifts ...
CPU time: 0.2360689640045166 seconds

> Top 10 drifts

('PROFESSION', 0.20745484400656822)
('MARITAL_STATUS', 0.14729533192587363)
('AGE', 0.10706075533661741)
('GENDER', 0.0280741262022205063)

> Deleted variables : []
> Drift coefficients dumped into directory : save
```

Optimiser 优化超参数:

```
opt = Optimiser(scoring = 'accuracy', n_folds = 3)
opt.evaluate(None, df)
```

以上代码输出如下:

```
No parameters set. Default configuration is tested
############################################# testing hyper-parameters... #############################################
>>> NA ENCODER :{'numerical_strategy': 'mean', 'categorical_strategy': '<NULL>'}

>>> CA ENCODER :{'strategy': 'label_encoding'}

>>> ESTIMATOR :{'strategy': 'LightGBM', 'boosting_type': 'gbdt', 'colsample_bytree': 0.8, 'learning_rate': 0.05, 'max_bin': 255, 'max_dept
h': -1, 'min_child_samples': 10, 'min_child_weight': 5, 'min_split_gain': 0, 'n_estimators': 500, 'nthread': -1, 'num_leaves': 31, 'object
ive': 'binary', 'reg_alpha': 0, 'reg_lambda': 0, 'seed': 0, 'silent': True, 'subsample': 0.9, 'subsample_for_bin': 50000, 'subsample_fre
q': 1}

MEAN SCORE : accuracy = 0.0570060141068
VARIANCE : 0.00237658458727 (fold 1 = 0.0558482613277, fold 2 = 0.0548523206751, fold 3 = 0.0603174603175)
CPU time: 87.5492041109924 seconds

0.057006014106759734
```

以下代码定义了机器学习流水线参数:

```
space = {
        'ne__numerical_strategy':{"search":"choice", "space":[0]},
        'ce__strategy':{"search":"choice",
                "space":["label_encoding","random_projection",
"entity_embedding"]},
        'fs__threshold':{"search":"uniform", "space":[0.01,0.3]},
        'est__max_depth':{"search":"choice", "space":[3,4,5,6,7]}
        }

best = opt.optimise(space, df,15)
```

以下输出显示用 LightGBM 机器学习算法对所选方法进行了测试:

也可以看到各种不同的度量指标，如准确度、方差和 CPU 时间：

调用 Predictor，用最佳模型做出预测：

```
predictor = Predictor()
predictor.fit_predict(best, df)
```

输出如下：

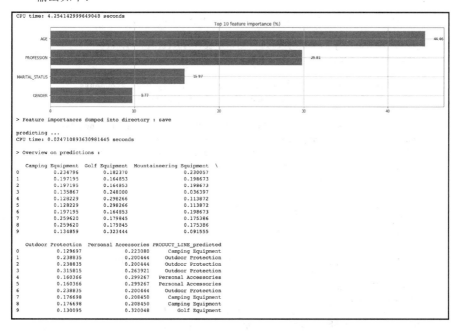

TPOT
TPOT

树型流水线优化工具（tree-based pipeline optimization tool，TPOT）采用基因编程发现性能最优的机器学习流水线，它是在 scikit-learn 上层构建的。

当数据集清洗并准备就绪时，TPOT 会辅助完成以下机器学习流水线的步骤。

- 特征预处理。
- 特征构建和选择。
- 模型选择。
- 超参数优化。

实验完成后，TPOT 就能提供最好的流水线。

TPOT 使用非常友好，如同用 scikit-learn 的 API 一般：

```
from tpot import TPOTClassifier
from sklearn.datasets import load_digits
from sklearn.model_selection import train_test_split

# Digits dataset that you have used in Auto-sklearn example
digits = load_digits()
X_train, X_test, y_train, y_test = train_test_split(digits.data,
digits.target,train_size=0.75,test_size=0.25)

# You will create your TPOT classifier with commonly used arguments
tpot = TPOTClassifier(generations=10, population_size=30, verbosity=2)

# When you invoke fit method, TPOT will create generations of populations,
# seeking best set of parameters. Arguments you have used to create
# TPOTClassifier such as generations and population_size will affect the
# search space and resulting pipeline.
tpot.fit(X_train, y_train)

print(tpot.score(X_test, y_test))
# 0.9834
tpot.export('my_pipeline.py')
```

将流水线导出为 my_pipeline.py 的 Python 文件后，可看到所选的流水线环节：

```
import numpy as np
import pandas as pd
from sklearn.model_selection import train_test_split
from sklearn.neighbors import KNeighborsClassifier

# NOTE: Make sure that the class is labeled 'target' in the data file
tpot_data = pd.read_csv('PATH/TO/DATA/FILE', sep='COLUMN_SEPARATOR',
dtype=np.float64)
features = tpot_data.drop('target', axis=1).values
training_features, testing_features, training_target, testing_target =\
train_test_split(features, tpot_data['target'].values,
random_state=42)
```

```
exported_pipeline = KNeighborsClassifier(n_neighbors=6,
weights="distance")

exported_pipeline.fit(training_features, training_target)
results = exported_pipeline.predict(testing_features)
```

搞定！

1.7 总结
Summary

到目前为止，你应该对什么是 AutoML 和为什么需要熟悉机器学习流水线有了基本的概念。

文中回顾了 AutoML 的核心环节，也练习使用了常用的 AutoML 库。

这当然不够，而且 AutoML 领域的研究活跃，不断有新库推出。建议也看看其他库，如 Auto-WEKA，该库在贝叶斯优化中采用了最新技术，以及 Xcessive，它是一个创建叠加集成模型的好工具。

好了，铺垫到此为止！接下来要开始着手打造自己的作品了，构建一个属于自己的 AutoML 系统！

第 2 章
Python 机器学习简介
Introduction to Machine Learning Using Python

第 1 章带你走进**机器学习**（machine learning，ML）的世界。本章会继续学习创建和使用**自动机器学习**（automated ML，AutoML）平台所必需的基础知识。很难每次都清楚地描述出如何应用机器学习最好或实现机器学习需要什么条件。但是，机器学习工具越来越易用，AutoML 平台也越来越接近更广泛的大众。未来无疑会有更多的人机协作。

机器学习的未来可能需要人来准备消耗数据和识别实现用例。更重要的是，需要人来解释机器学习结果和审计机器学习系统，看系统是否按照最合适或最好的方式解决问题。未来看似精彩，但需要我们构建。这也是本书中要解决的问题。本章会覆盖以下主题。

- 机器学习流程及其不同类型。
- 监督学习——回归和分类。
- 无监督学习——聚类。
- 集成学习——套袋、提升和叠加。
- 根据数据推断的任务。
- 特定任务评估指标。

要清楚，只靠一个章节是不足以学会和练习好机器学习的。关于机器学习，市面上已经有很多优质的书籍和材料了，关于上述主题都可在其中找到详细的内容。书后附录中的其他推荐书籍也可找到一些参考材料。本章的目标是纵览各种不同的机器学习技术，讨论后续章节工作所必需的一些内容。

因此，机器是乐于学习的。你愿意帮助它们吗？抓紧时间。首先看一下机器学习是什么！

2.1 技术要求
Technical Requirements

全部的代码示例可在 GitHub 的 `Chapter 02` 文件夹中找到。

2.2 机器学习
Machine Learning

机器学习要追溯到几世纪以前。理论依据是计算机不用编程就可以学着完成任务。机器学习的迭代也是举足轻重的一方面，因为机器要始终不断地适应新数据。机器需要从历史数据中学习，优化实现更好的计算，也需要泛化能力以产出合理的成果。

大家都注意到了以前的系统主要是基于规则的，即我们准备一套预定义条件，让机器执行并产生所需结果。如果机器可以自己学习这些规则，分发结果，并能解释所发现的结果，那该多好呀！这就是机器学习。机器学习是一个广义的概念，包括机器学习数据要用到的各种方法和算法。作为**人工智能**（artificial intelligence，AI）的一个分支，机器学习算法经常用来发现隐藏的模式，建立关联关系，也会用于预测结果。

机器学习依赖一些格式化的输入，根据任务输出结果。输入格式是根据所用机器学习类型和所考虑的算法而定的。这种输入数据的具体表示就叫**特征**（features）或**预测器**（predictors）。

机器学习流程
Machine Learning process

我们自己如何学习？在学校学习时，是靠老师教。我们从老师的教导中学习（训练）。在学期末，需要参加考试（测试），基本就是验证所学到的知识。所取得的分数决定了我们的命运（评估）。通常评估要确定一个及格线（基线）。分数决定了我们是需要重考这一科还是可以继续深造（部署）。

机器也是这样学习的。括号中的词语就是机器学习专业人士所用的术语。然而，这只是人类和机器开展学习的方法之一。这是典型的监督学习方法。人们有时也从经验中学习，这是无监督学习。下面我们一起了解这些学习方法的更多细节。

广义上，如上所述，机器学习算法分两大类：监督学习和无监督学习。但还有一些其他类型的机器学习，如强化学习、迁移学习和半监督学习，但由于用得较少，本书中不做讨论。

监督学习
Supervised Learning

顾名思义，学习过程是根据一个特定的目标或结果受到监督的。

监督机器学习模型的目标是学习和发现可以正常预测结果的模式。在监督学习中，总会有一个带目标属性标签的历史数据集。除目标之外的其他属性均被称为预测器/特征。

目标可以是一个连续的数值属性，也可以是决定是或否的二值属性，还可以是有超过两个结果的多类属性。根据这个目标，模型会识别出一个模式，在预测器之间建立关系，然后用学习出来的模型在全新的独立数据集中预测未知目标。

许多机器学习算法都归类为监督学习法，简单列举如线性回归、逻辑回归、决策树、随机森林和**支持向量机**（support vector machines，SVMs）等。

在机器学习项目中,最艰巨的工作就是为一个任务识别和选出最合适的算法。这也是创建 AutoML 系统时需要重点关注的一个环节。算法选择的过程是由多种因素决定的,本书会详细讲解。

无监督学习
Unsupervised Learning

以此类推,无监督学习中是没有目标属性的。无监督学习的目的是从输入数据集的特征结构和关系中推导和识别出模式。无监督学习可用于发现定义群组的规则,如主题提取;可用于拆分,如客户细分;或可用于确定数据内部结构,如基因聚类。无监督学习算法的示例包括关联规则挖掘和聚类算法。

创建 AutoML 系统前,有必要了解一下不同的学习算法。采用一种算法,关键是了解 3W,即 What(是什么)、Where(在哪里使用)和 What method(用什么方法实现)。

接下来会找出不同算法的 3W,这有助于创建一个可靠的 AutoML 系统。

2.3 线性回归
Linear Regression

首先从线性回归的 3W 开始。

什么是线性回归
What is Linear Regression

这是一种最常用的传统回归分析。线性回归研究非常严谨,并广泛应用于实践中。线性回归是确定一个因变量(y)和一个或多个自变量(x)之间关系的方法。利用推演出的关系,可从观测值 x 中预测出一个未知的 y。用数学公式表达的话,如果 x 是一个自变量(通常称为预测器),y 是因变量(也称目标变量),那么它们的关系表示如下:

$$y = mx + b + \epsilon$$

其中：m 表示直线的斜率；b 表示最佳拟合回归线的截距；ϵ 表示误差项，误差是实际值和预测值之间的偏差。

这是一个很简单的线性回归公式，因为只有一个预测器（x）和一个目标（y）。用多个预测器预测一个目标时，称为**多元线性回归**（multiple linear regression）。术语**线性**（linear）含有基础假设，即标的数据之间展现出线性关系。

我们来构建两个变量之间的散点图，两个变量分别为一种产品的**销量**（quantity sold）和**收入**（revenue）。从图 2-1 中可推导出两个变量是正相关的，即当产品销量上涨时，收入也增高。但是，我们却无法量化这种关系，即无法通过销量预测收入。

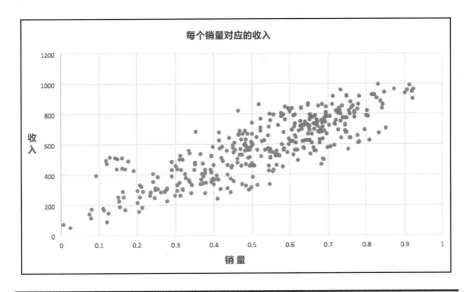

图 2-1　销量收入

如果将图 2-1 适当延伸，加一条趋势线，则可看到最佳拟合线，如图 2-2 所示。在这条线上的任何数据点都完美地预测了收入值。远离这条线时，预测的可靠性就降低了。

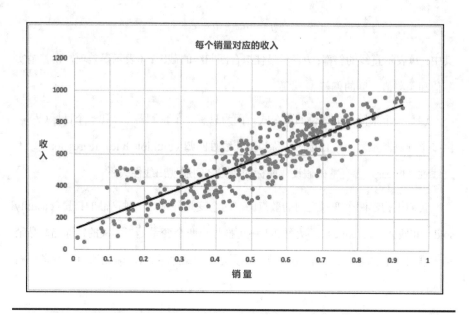

图 2-2 销量收入带趋势线

那么，怎么找到最佳拟合线呢？最常见和广泛使用的技术是**普通最小二乘法**（ordinary least square，OLS）估算。

OLS 回归的工作原理

OLS `LinearRegression` 线性回归方法是寻找数据的拟合函数最直接的方法。通过最小化数据的**误差平方和**（sum of squared errors，SSE），找出最佳拟合线。SSE 是实际值与平均值偏差的和。然而，简单往往是有代价的。一种好的 OLS 方法要付出的代价就是必须遵守几个基本假设。

OLS 的假设

只有当这些数据假设全部成立时，OLS 回归技术的优势才能体现出来。

- **线性**（linearity）：X 和 Y 之间的真正关联关系是线性的。
- **同方差**（homoscedastic）：残差的方差必须是常数。残差是目标观测值与预测值之间的差。
- **正态**（normality）：残差/误差应该是正态分布的。

- 没有或只有较少的多重共线性（no or little multicollinearity）：残差/误差必须是独立变量。

数据中有异常值也会影响 OLS。用 OLS 进行线性回归建模时，必须进行异常值处理。

在哪里用线性回归
Where is Linear Regression Used?

线性回归有许多实际用例，多数可以归结为以下两大类。

- 如果目标是预测或预报，可用线性回归对由因变量和自变量组成的已知数据集构建预测模型。
- 如果目标是确定目标和预测值之间的关系强度，则可用线性回归对已知X值对应Y值的变化进行量化。

采用什么方法实现线性回归
By Which Method Can Linear Regression Be Implemented?

可以用 scikit-learn 的 `LinearRegresssion` 方法创建一个 Python 的线性回归模型。由于这是第一个用 Python 实现模型的例子，所以我们先开个小差，了解一些用 Python 构建模型的必备软件包。

- `numpy`：这是数学函数中用到的数值计算 Python 模块。其中有健壮的数据结构，能确保多维数组和矩阵的计算有效性。
- `pandas`：为数据修改提供 DataFrame 对象。DataFrame 可容纳不同类型的值和数组。padnas 使用 Python 读取、写入和处理数据。
- `scikit-learn`：这是一个 Python 的机器学习库，包括许多机器学习算法，是用 Python 构建机器学习模型很常用的一个库。除了机器学习算法，还提供开发模型所必需的其他各种函数，如 `train_test_split`，还有模型评估指标和优化指标。

构建模型前，首先要将这些必备库导入 Python 环境中。如果在 Jupyter Notebook 中运行代码，则有必要声明 `%matplotlib inline`，以便在线查看图表。我们需要导入 `numpy` 和 `pandas` 包，便于数据修改和数值计算。本练习的目标是创建一个线性回归模型，因此也需要从 scikit-learn 包中导入 `LinearRegression` 方法。本任务中使用 scikit-learn 的 `Boston` 示例数据集：

```
%matplotlib inline
import numpy as np
import pandas as pd
from sklearn.linear_model import LinearRegression
import matplotlib.pyplot as plt
from sklearn.datasets import load_boston
```

接着，用以下命令加载 Boston 数据集。这是一个字典，可使用键（keys）查看内容：

```
boston_data = load_boston()
boston_data.keys()
```

以上代码输出如下：

```
Out[2]: dict_keys(['data', 'target', 'feature_names', 'DESCR'])
```

`boston_data` 有 4 个键，对其所指代的值是自说明的。可以从 `data` 和 `target` 键中提取到数据和目标值。`feature_names` 键对应属性名称，`DESCR` 是对每种属性的描述。

处理数据前先看看数据大小是一个好习惯。从而决定是直接用全部数据，还是只取其中一部分样本，也能推测执行时间大概需要多久。

使用 Python 的 `data.shape` 函数查看数据维度（行和列）非常方便。

```
print(" Number of rows and columns in the data set ",
boston_data.data.shape)
print(boston_data.feature_names)
```

以上代码输出如下：

```
Number of rows and columns in the data set  (506, 13)
['CRIM' 'ZN' 'INDUS' 'CHAS' 'NOX' 'RM' 'AGE' 'DIS' 'RAD' 'TAX' 'PTRATIO'
 'B' 'LSTAT']
```

接下来，要将字典转换为 `DataFrame`。调用 `pandas` 库中的 `DataFrame` 函数即可完成。使用 `head()` 显示记录子集，验证数据：

```
boston_df =pd.DataFrame(boston_data.data)
boston_df.head()
```

 DataFrame 是一个向量集，可当成二维表格处理。可以把 DataFrame 每行对应一个观测值，每列对应该观测值的属性。这在机器学习建模拟合时特别有用。

以上代码输出如图 2-3 所示。

Out[4]:		0	1	2	3	4	5	6	7	8	9	10	11	12
	0	0.00632	18.0	2.31	0.0	0.538	6.575	65.2	4.0900	1.0	296.0	15.3	396.90	4.98
	1	0.02731	0.0	7.07	0.0	0.469	6.421	78.9	4.9671	2.0	242.0	17.8	396.90	9.14
	2	0.02729	0.0	7.07	0.0	0.469	7.185	61.1	4.9671	2.0	242.0	17.8	392.83	4.03
	3	0.03237	0.0	2.18	0.0	0.458	6.998	45.8	6.0622	3.0	222.0	18.7	394.63	2.94
	4	0.06905	0.0	2.18	0.0	0.458	7.147	54.2	6.0622	3.0	222.0	18.7	396.90	5.33

图 2-3　DataFrame 转换

可见列名只是数值序号，并没有给出 DataFrame 的含义。因此，就把 feature_names 作为列名分配给 boston_df，获得有意义的名称：

```
boston_df.columns = boston_data.feature_names
```

再次查看 boston 房屋租金数据样本，列描述清晰多了：

```
boston_df.head()
```

以上代码输出如图 2-4 所示。

Out[5]:		CRIM	ZN	INDUS	CHAS	NOX	RM	AGE	DIS	RAD	TAX	PTRATIO	B	LSTAT
	0	0.00632	18.0	2.31	0.0	0.538	6.575	65.2	4.0900	1.0	296.0	15.3	396.90	4.98
	1	0.02731	0.0	7.07	0.0	0.469	6.421	78.9	4.9671	2.0	242.0	17.8	396.90	9.14
	2	0.02729	0.0	7.07	0.0	0.469	7.185	61.1	4.9671	2.0	242.0	17.8	392.83	4.03
	3	0.03237	0.0	2.18	0.0	0.458	6.998	45.8	6.0622	3.0	222.0	18.7	394.63	2.94
	4	0.06905	0.0	2.18	0.0	0.458	7.147	54.2	6.0622	3.0	222.0	18.7	396.90	5.33

图 2-4　赋予列名

在线性回归中，必须有一个 DataFrame 是目标变量，另一个带有其他特征的 DataFrame 作为预测器。本练习的目标是预测房价，所以把 PRICE 当作目标属性

（Y），其他的都是预测器（x）。使用 drop 函数将 PRICE 从预测器列表中去掉。

下一步，打印每个变量的截距和系数。系数决定每个预测器对预测房价（目标 Y）的权重和影响。截距是一个常量，在一个预测器都没有的情况下，这个常量就可视为房价：

```
boston_df['PRICE'] = boston_data.target
X = boston_df.drop('PRICE', axis=1)
lm = LinearRegression()
lm.fit(X, boston_df.PRICE)
print("Intercept: ", lm.intercept_)
print("Coefficient: ", lm.coef_)
```

以上代码输出如下：

```
Intercept: 36.4911032804
Coefficient: [ -1.07170557e-01   4.63952195e-02   2.08602395e-02   2.68856140e+00
 -1.77957587e+01   3.80475246e+00   7.51061703e-04  -1.47575880e+00
  3.05655038e-01  -1.23293463e-02  -9.53463555e-01   9.39251272e-03
 -5.25466633e-01]
```

之前的截图中看不出哪个系数对应哪个预测器。所以，使用以下代码绑定特征和系数：

```
pd.DataFrame(list(zip(X.columns, lm.coef_)),columns=
['features','estimatedCoefficients'])
```

以上代码输出如图 2-5 所示。

	features	estimatedCoefficients
0	CRIM	-0.107171
1	ZN	0.046395
2	INDUS	0.020860
3	CHAS	2.688561
4	NOX	-17.795759
5	RM	3.804752
6	AGE	0.000751
7	DIS	-1.475759
8	RAD	0.305655
9	TAX	-0.012329
10	PTRATIO	-0.953464
11	B	0.009393
12	LSTAT	-0.525467

图 2-5　特征和系统绑定

接下来计算和查看均方误差指标。现在假设这是模型预测房价时的平均误差。评估指标对理解模型动态以及模型在生产环境中的性能如何都有很重要的意义：

```
lm.predict(X)[0:5]
mseFull = np.mean((boston_df.PRICE - lm.predict(X)) ** 2)
print(mseFull)
```

上述代码输出如下：

```
21.897779217687486
```

模型是在整个数据集上构建的，但必须保证所开发的模型在真实生产环境中适用于不同的数据集。因此，建模用的数据一般按 70:30 的比例拆分成两个子集。大的子集用于训练模型，另一个子集则用来测试模型。这个独立的测试子数据集被视为伪生产环境（dummy production environment），因为在模型训练期间它是隐藏的。测试子数据集用于生成预测和评估模型的准确性。scikit-learn 提供了一种可以拆分数据集的方法 train_test_split。函数中 test_size 参数表示百分之多少的数据会留作测试数据。在如下代码中，我们将数据集分成 train 子集和 test 子集，然后重新训练模型：

```
#Train and Test set
from sklearn.model_selection import train_test_split
X_train, X_test, Y_train, Y_test = train_test_split(X, boston_df.PRICE,
test_size=0.3, random_state=42)
print(X_train)
```

由于设置了 test_size=0.3，因此 70%的数据集会用来创建 train 数据集，30%的数据集会用来创建 test 数据集。同之前的步骤，创建一个线性回归模型，但现在只用（X_train 和 Y_train）训练数据集创建模型：

```
lm_tts = LinearRegression()
lm_tts.fit(X_train, Y_train)
print("Intercept: ", lm_tts.intercept_)
print("Coefficient: ", lm_tts.coef_)
```

上述代码输出如下：

```
Intercept: 31.6821485821
Coefficient: [ -1.32774155e-01   3.57812335e-02   4.99454423e-02   3.12127706e+00
 -1.54698463e+01   4.04872721e+00  -1.07515901e-02  -1.38699758e+00
  2.42353741e-01  -8.69095363e-03  -9.11917342e-01   1.19435253e-02
 -5.48080157e-01]
```

为 train 数据集和 test 数据集分别预测目标值，计算其**均方误差**（mean squared error，MSE）：

```
pred_train = lm.predict(X_train)
pred_test = lm.predict(X_test)
print("MSE for Y_train:", np.mean((Y_train - lm.predict(X_train)) ** 2))
print("MSE with Y_test:", np.mean((Y_test - lm.predict(X_test)) ** 2))
```

以上代码输出如下：

```
MSE for Y_train: 22.86266796675359
MSE with Y_test: 19.650604104730895
```

可得 train 数据集和 test 数据集的 MSE 分别为 22.86 和 19.65。这说明此模型的性能在训练和测试阶段很相近，可在另一个相同的新数据集中部署以预测房价。

接下来画一个残差曲线图，看残差是否遵循线性模式：

```
plt.scatter(pred_train,pred_train - Y_train, c = 'b',s=40,alpha=0.5)
plt.scatter(pred_test,pred_test - Y_test, c = 'r',s=40,alpha=0.7)
plt.hlines(y = 0, xmin=0, xmax = 50)
plt.title('Residual Plot - training data (blue) and test data(green)')
plt.ylabel('Residuals')
```

以上代码输出如图 2-6 所示。

残差对称地分布在水平线周围，也就是说，确实展现出了完美的线性模式。

开发一个模型很容易，但设计一个有用的模型则很难。评估机器学习模型性能是在机器学习流水线中非常关键的一步。模型建好后，必须评估并确定模型的准确性。在后面的章节中会详细介绍评估回归模型时常用的一些评估指标。

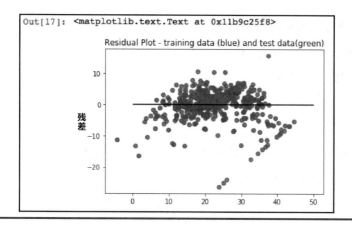

图 2-6 残差图——训练数据（蓝）和测试数据（绿）

2.4 重要评估指标——回归算法
Important Evaluation Metrics -Regression Algorithms

评估机器学习模型的过程分为两个阶段。第一阶段，必须评估模型统计学意义上的准确性，即统计学假设是否正确、模型性能是否突出、其他独立的数据集中是否也能实现同样的性能。要完成这些评估，需要几个模型评估指标。第二阶段，评估模型的结果是否满足业务要求，业主是否确实能从中得到一些启发或有用的预测。

评估回归模型会使用以下指标：

平均绝对误差（mean absolute error，MAE）：预测误差的绝对值之和。预测误差是指预测值和实际值之间的差异。这个指标可表示误差的量级。但是，还无法判断模型是否过度预测或预测不足。MAE 分值越低越好：

$$MAE = \sum_{i=1}^{n} \frac{|y_i - y_i^{\wedge}|}{n}$$

其中，y_i=实际值；$y_i^{\wedge}$=预测值；n=用例（记录）数量。

均方误差（mean squared error，MSE）：方差和的平均值。本指标表示误差的量级和方向。但是，度量单位因平方值而发生了变化。这个不足由另一个指标补齐：均方根误差。值越小，模型越好：

$$MSE = \sum_{i=1}^{n} \frac{(y_i - y_i^A)^2}{n}$$

均方根误差（root mean square error，RMSE）：均方差的平方根。取平方根之后，度量单位就恢复到原来的单位。RMSE 分越低，模型越好：

$$RMSE = \sqrt{\sum_{i=1}^{n} \frac{(y_i - y_i^A)^2}{n}}$$

R^2 分：也称**确定系数**（coefficient of determination）。表示模型所解释的方差的百分比。比如，若 R^2 是 0.9，则模型中所用到的属性或特征代表方差的 90%。R^2 值分布在 0 到 1 之间，值越高，模型质量越好。但是，要设定好测试策略，验证模型没有过拟合：

$$R^2 = 1 - \sum_{i=1}^{n} \frac{(y_i - y_i^A)^2}{(y_i - y_i^-)^2}$$

其中：y_i=实际值；y_i^A=预测值；n=用例（记录）数量；y_i^-=y的均值。

> 当机器学习模型学习训练数据太过于好时，就会发生过拟合。模型结果偏差低，方差高。这样的模型可能在预测新数据时表现很差。

本节我们学了回归分析，监督机器学习法的一种。适用于目标数据是连续数值型数据的场景，如预测员工期望薪资、预测房价或预测花费。

如果目标数据是离散的呢？如何预测一个客户是否可能流失？如何预测是否应该给一个潜在用户批准贷款/信用卡？这些情况下线性回归不再适用，因为这些问题违背了线性回归的假设前提。那有其他办法吗？有，可以使用分类模型。

> 分类模型是另一种监督学习法，可预测离散输入数据的目标值。分类算法也称**分类器**（classifiers），可识别出输入数据所支持的类别，利用这些信息给未识别的或未知目标标签分配一个类。

接下来将会学习一些常用的分类器，如逻辑回归、决策树、SVM 和 KNN。逻辑回归可看成是回归和分类法之间的桥梁。它是一种隐藏在特征回归中的分类器。但是，这是最有效且最易解释的分类模型之一。

2.5 逻辑回归
Logistic Regression

我们从逻辑回归的3W开始。先来回顾一下3W法，首先要问What（是什么），然后是Where（在哪里使用），最后是What method（用什么方法实现）。

什么是逻辑回归
What Is Logistic Regression?

逻辑回归可算作是线性回归算法的延伸。基本上与线性回归相同，但主要用于离散或分类结果。

在哪里使用逻辑回归
Where Is Logistic Regression Used?

逻辑回归应用在离散目标变量场景下，如二分响应。在这些场景下，一些线性回归的假设，如目标属性和特征，不再是线性关系，残差也可能不是正态分布，或者误差项是异方差的。在逻辑回归中，将目标重构为其优势比的对数，再拟合回归方程，如下所示：

$$log(\frac{P}{P-1}) = mx + b + \epsilon$$

优势比反映了某个事件发生的概率或可能性与其不发生的概率之比。若P是一个事件/类别存在的概率，则$P-1$是另一个事件/类存在的概率。

使用什么方法实现逻辑回归
By which method can logistic regression be implemented?

导入 scikit-learn 的 `LogisticRegression` 方法，可构建逻辑回归模型。使用此前

创建线性回归模型时的方法加载所需包：

```python
import pandas as pd
import numpy as np
from sklearn import preprocessing
import matplotlib.pyplot as plt
from sklearn.linear_model import LogisticRegression
```

这次使用人力部门的数据集，其中包括一个清单，列出了谁已经离职，谁仍在这个岗位上：

```python
hr_data = pd.read_csv('data/hr.csv', header=0)
hr_data.head()
hr_data = hr_data.dropna()
print(hr_data.shape)
print(list(hr_data.columns))
```

以上代码输出如下：

```
(14999, 10)
['satisfaction_level', 'last_evaluation', 'number_project', 'average_montly_hours', 'time_spend_company', 'Work_accident', 'left', 'promotion_last_5years', 'sales', 'salary']
```

数据集中有 14999 行和 10 列。data.columns 函数显示了属性名称。salary（薪资）属性有三个值：high（高）、low（低）和 medium（中），而且 sales（销售）有七个值：IT、RandD（研发）、marketing（市场）、product_mng（产品管理）、sales（销售）、support（支持）和 technical（技术）。这些离散的数据要应用到模型中，需先转换成数值格式。转换方法有很多种。其中一种是伪编码，也称**独热编码**（one-hot encoding）。使用这种方法，可以给每一个类别属性都赋一个伪列。

对每个伪属性而言，1 代表这个类存在，0 代表不存在。

> 离散数据可以是标称的或序数的。当离散数据中有一套自然排序的值时，称为**序数**(ordinal)。如 high、medium 和 low 的类别值就是序数值。在这种情况中大多数会使用标签编码。若在类别或离散值中无法推演出任何关系或顺序，就叫**标称**（nominal）。如红、黄、绿等颜色就没有顺序。这时常用伪编码。

pandas 中的 get_dummies 方法提供了使用 Python 创建伪变量的易用接口。函数的输入值为数据集和需要伪编码的属性名称。本例中，要对人力数据集中的 salary 和 sales 属性进行伪编码：

```python
data_trnsf = pd.get_dummies(hr_data, columns =['salary', 'sales'])
data_trnsf.columns
```

以上代码输出如下：

```
Index(['satisfaction_level', 'last_evaluation', 'number_project',
       'average_montly_hours', 'time_spend_company', 'Work_accident', 'left',
       'promotion_last_5years', 'salary_high', 'salary_low', 'salary_medium',
       'sales_IT', 'sales_RandD', 'sales_accounting', 'sales_hr',
       'sales_management', 'sales_marketing', 'sales_product_mng',
       'sales_sales', 'sales_support', 'sales_technical'],
      dtype='object')
```

这时可以对数据集建模了。sales 和 salary 属性成功完成了独热编码。下一步，预测人员流失，需要把 left（离职）属性作为目标，因为其中包含员工是否离职的信息。可从代码中使用 X 表示输入预测器数据集中去掉 left 后的数据。left 属性由 Y（目标）表示：

```
X = data_trnsf.drop('left', axis=1)
X.columns
```

以上代码输出如下：

```
Index(['satisfaction_level', 'last_evaluation', 'number_project',
       'average_montly_hours', 'time_spend_company', 'Work_accident',
       'promotion_last_5years', 'salary_high', 'salary_low', 'salary_medium',
       'sales_IT', 'sales_RandD', 'sales_accounting', 'sales_hr',
       'sales_management', 'sales_marketing', 'sales_product_mng',
       'sales_sales', 'sales_support', 'sales_technical'],
      dtype='object')
```

把数据集按 70:30 分成 train 子集和 test 子集。70%的数据用于训练逻辑回归模型，余下 30%的数据用于评估模型准确性：

```
from sklearn.model_selection import train_test_split
X_train, X_test, Y_train, Y_test = train_test_split(X, data_trnsf.left,
test_size=0.3, random_state=42)
print(X_train)
```

运行以上代码片段，生成 4 个数据集。X_train 和 X_test 是 train 和 test 的预测器数据集。Y_train 和 Y_test 是 train 和 test 的目标数据集。现在用模型拟合训练数据，在测试数据中评估模型的准确性。首先，创建一个 LogisticsRegression() 分类器类实例。然后，用分类器拟合训练数据：

```
attrition_classifier = LogisticRegression()
attrition_classifier.fit(X_train, Y_train)
```

模型构建成功后，可使用 predict 方法在 X_test 测试预测器数据集中预测对应的目标值（Y_pred）：

```
Y_pred = attrition_classifier.predict(X_test)
```

创建一个 confusion_matrix 评估分类器。多数模型评估指标都是基于混淆矩阵本身的。稍后会详细讨论混淆矩阵和其他评估指标。目前先把混淆矩阵看成是 4 个值的矩阵，表示预测正确和预测不正确的值的数量。根据混淆矩阵的值，可计算出分类器的准确性。本例中的分类器准确性为 0.79 或 79%，即 79%的用例是预测正确的：

```
from sklearn.metrics import confusion_matrix
confusion_matrix = confusion_matrix(Y_test, Y_pred)
print(confusion_matrix)

print('Accuracy of logistic regression model on test dataset:
{:.2f}'.format(attrition_classifier.score(X_test, Y_test)))
```

以上代码输出如下：

```
[[3175  253]
 [ 711  361]]
Accuracy of logistic regression classifier on test set: 0.79
```

有时，准确率不一定是判断模型性能的好标准。比如，在不均衡的数据集中，预测可能会偏向数量更多的类。因此，要看看其他指标，比如 F1 得分、**曲线下面积**（area under curve，AUC）、精度、召回等，这样可对模型做出更合理的判断。从 scikit-learn 的 metric 方法中导入 classification_report，可获得全部指标的分值：

```
from sklearn.metrics import classification_report
print(classification_report(Y_test, Y_pred))
```

以上代码输出如下：

```
             precision    recall  f1-score   support

          0       0.82      0.93      0.87      3428
          1       0.59      0.34      0.43      1072

avg / total       0.76      0.79      0.76      4500
```

受试者操作特性（receiver operating charateristic，ROC）是视觉化二进制分类器性能中最常用的方法。AUC 是指 ROC 曲线下的面积，在 ROC 曲线的基础上总结出分类器性能的一个数值。使用以下的 Python 代码可画一条 ROC 曲线：

```python
from sklearn.metrics import roc_curve
from sklearn.metrics import auc

# Compute false positive rate(fpr), true positive rate(tpr), thresholds and
roc auc(Area under Curve)
fpr, tpr, thresholds = roc_curve(Y_test, Y_pred)
auc = auc(fpr,tpr)

# Plot ROC curve
plt.plot(fpr, tpr, label='AUC = %0.2f' % auc)
#random prediction curve
plt.plot([0, 1], [0, 1], 'k--')
#Set the x limits
plt.xlim([0.0, 1.0])
#Set the Y limits
plt.ylim([0.0, 1.0])
#Set the X label
plt.xlabel('False Positive Rate(FPR) ')
#Set the Y label
plt.ylabel('True Positive Rate(TPR)')
#Set the plot title
plt.title('Receiver Operating Characteristic(ROC) Cure')
# Location of the AUC legend
plt.legend(loc="right")
```

以上代码输出如图 2-7 所示。

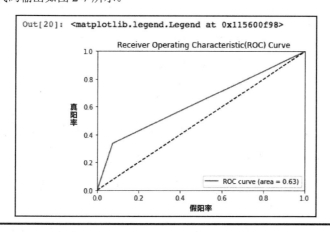

图 2-7　受试者操作特性（ROC）曲线

本例模型的 AUC 是 0.63。我们已经了解过一些评估分类模型的指标，其中有些比较陌生。所以，先了解下这些指标，再继续讨论分类算法。

2.6 重要评估指标——分类算法
Important Evaluation Metrics Classification Algorithms

多数用来评估分类模型的指标是以从混淆矩阵四象限中得到的值为依据的。从本节开始，先了解这些指标是什么。

- **混淆矩阵**（confusion matrix）：这是评估分类模型（即分类器）的里程碑。顾名思义，这个矩阵有时是混淆的。假设把混淆矩阵看成是一个图表中的两个轴。X 轴标签是预测，有两个值，即阳性（Positive）和阴性（Negative）。相应地，Y 轴标签也有相同的两个值，即阳性（Positive）和阴性（Negative），如图 2-8 所示。这个矩阵是一个表格，包含分类器的实际值和预测值的数量。

图 2-8　混淆矩阵

- 如果要推演矩阵中每个象限的信息。
    - 象限一，也称真阳性（true positive，TP），是准确识别出来的阳性类预测的数量；
    - 象限二，也称假阳性（false positive，FP），是实际阳性类用例预测错误的数量；
    - 象限三，即假阴性（false negative，FN），是指阴性类用例预测错误的数量；
    - 象限四，即真阴性（true negative，N），是准确分类出来的阴性类预测的数量。

- 准确率（accuracy）：准确率用来衡量分类器做出准确预测的频率，是正确预测数量占总预测数量的比例：

$$Accuracy = TP/(TP + FP + FN + TN)$$

- 精度（precision）：精度用于评估准确识别出来的真阳性所占的比例，是真阳性在全部阳性预测中的占比：

$$Precision = TP/(TP + FN)$$

- 召回（recall）：也称灵敏度或真阳率（true positive rate，TPR）。估算的是真阳性占观测到的全部目标阳性值的比例：

$$Recall = TP/(TP + FP)$$

- 错分类率（misscalssification rate）：表示分类器预测不准确的频率，是不正确预测占全部预测的比例：

$$Misclassification\ Rate = (FP + FN)/(TP + FP + FN + TN)$$

- 特异性（specificity）：特异性又称真阴率（true negative rate，TNR），表示真阴性占目标全部观察到的阴性值的比例：

$$Specificity = TN/(FP + TN)$$

- ROC 曲线（ROC curve）：ROC 曲线总结了分类器在全部可能的阈值中的性能。ROC 曲线图表是全部可能的阈值的分布图，y 轴是真阳率（TPR），x 轴是假阳率（FPR）。

- AUC：AUC 是 ROC 曲线下的面积。如果分类器出色，真阳率就会升高，曲线下的面积就会趋近于 1。若分类器与随机猜测相似，则真阳率会与假阳率（1-灵敏度）呈线性增长。本例中，AUC 约为 0.5。AUC 值越好，模型质量越高。

- 提升指数（lift）：提升指数是估算模型相对于平均或基线模型的预测能力的改善程度。比如，人力流失数据集的基线模型是 40%，但同一数据集上新模型的准确率却为 80%，那么表示模型提升指数到了 2（80/40）。

- **均衡准确率**（balanced accuracy）：有时，准确率不足以评估模型的好坏。在不均衡的数据集中，准确率就可能不是很有效。这种情况下，可用均衡准确率作为评估指标之一。均衡准确率是从任意类中得到的平均准确率计算出来的度量值：

$$Balanced\ Accuracy = 0.5 * ((TP/(TP + FN)) + (TN/(TN + FP))$$

> 不均衡数据集：当一个类的数量远多于另一个类的数量时，这种情况下，存在固有偏差会导致预测倾向数量大的类。但是，基础学习器，如决策树和逻辑回归，也有这个问题。集成模型，如随机森林，则可很好地处理不均衡类。

- **F1 得分**（F1-score）：F1 得分也是一种评估不均衡分类器的好方法。F1 得分是精度和召回的调和中项。取值在 0 到 1 之间：

$$F1\ Score = 2 * PR/(P + R), where\ P = Precision, R = Recall$$

- **汉明损失**（Hamming loss）：识别预测错误的那部分标签。

- **马修斯关联系数**（Matthews correlation coefficient，MCC）：MCC 是目标和预测之间的关联系数。取值在−1 和+1 之间。实际和预测之间完全不一致时为−1，实际和预测之间完全吻合时为 1，预测与实际之间随机无关时为 0。由于涉及混淆矩阵中全部 4 个象限的值，通常视此系统为均衡度量指标。

有时，为预测构建模型不只是一个要求。我们要思考如何构建模型，要关注描述模型的关键特征。决策树就是一种满足此条件的理想模型。

2.7 决策树
Decision Trees

决策树有一套透明的表示规则可促进分类/预测，得益于这种透明性，决策树成为机器学习世界中广泛使用的分类器。我们通过 3W 问题进一步了解决策树算法。

什么是决策树
What are Decision Trees?

决策树是一种分层次的树状结构,很容易说明和解释。这种算法不易受异常值影响。构建决策树的过程是一种递归划分方法,将训练数据分成不同的组,目的是找出同质纯子组,即只有一个类的数据。

 异常值与其他数据点距离非常远,会扭曲数据的分布。

在哪里使用决策树
Where are Decision Trees used?

决策树很适合用来解释为什么做出某个决策。比如,财政机构在批准一笔贷款或下发信用卡之前,可能需要完整地描述出哪些规则影响顾客的信用评分。

使用什么方法实现决策树
By which method can decision trees be implemented?

通过导入 scikit-learn 的 `DecisionTreeClassifier` 可构建决策树模型:

```
import numpy as np
import pandas as pd
from sklearn.tree import DecisionTreeClassifier
from sklearn.metrics import accuracy_score
from sklearn import tree
```

下一步,读取人力流失数据集,按照在逻辑回归示例中的步骤进行一遍数据预处理:

```
hr_data = pd.read_csv('data/hr.csv', header=0)
hr_data.head()
hr_data = hr_data.dropna()
print(" Data Set Shape ", hr_data.shape)
print(list(hr_data.columns))
print(" Sample Data ", hr_data.head())
```

以上代码输出如下:

```
Data Set Shape  (14999, 10)
['satisfaction_level', 'last_evaluation', 'number_project', 'average_montly_hours', 'time_spend_company', 'Work_accid
ent', 'left', 'promotion_last_5years', 'sales', 'salary']
 Sample Data    satisfaction_level  last_evaluation  number_project  average_montly_hours  \
0                      0.38             0.53               2                 157
1                      0.80             0.86               5                 262
2                      0.11             0.88               7                 272
3                      0.72             0.87               5                 223
4                      0.37             0.52               2                 159

   time_spend_company  Work_accident  left  promotion_last_5years  sales  \
0          3                  0          1             0           sales
1          6                  0          1             0           sales
2          4                  0          1             0           sales
3          5                  0          1             0           sales
4          3                  0          1             0           sales

   salary
0    low
1   medium
2   medium
3    low
4    low
```

以下代码会为类别型数据创建伪变量，并把数据集分成 train 子集和 test 子集：

```
data_trnsf = pd.get_dummies(hr_data, columns =['salary', 'sales'])
data_trnsf.columns
X = data_trnsf.drop('left', axis=1)
X.columns
from sklearn.model_selection import train_test_split
X_train, X_test, Y_train, Y_test = train_test_split(X, data_trnsf.left,
test_size=0.3, random_state=42)
print(X_train)
```

接下来，将 DecisionTreeClassifier 实例化，至少要包含必选参数，构建决策树分类器。以下是一些用来生成决策树模型的参数。

- criterion：形成决策树的不纯指标；可以是 entropy 或 gini。
- max_depth：最大树深。
- min_sample_leafs：构建一个叶子节点所需的最少样本数量。
- max_depth 和 min_sample_leafs 是树的两个预修剪标准。

使用这些参数构建一个决策树模型：

```
attrition_tree = DecisionTreeClassifier(criterion = "gini", random_state = 100,
max_depth=3, min_samples_leaf=5)
attrition_tree.fit(X_train, Y_train)
```

以上代码输出如下：

```
Out[8]: DecisionTreeClassifier(class_weight=None, criterion='gini', max_depth=3,
            max_features=None, max_leaf_nodes=None,
            min_impurity_split=1e-07, min_samples_leaf=5,
            min_samples_split=2, min_weight_fraction_leaf=0.0,
            presort=False, random_state=100, splitter='best')
```

接着，生成一个混淆矩阵评估模型，代码如下：

```
Y_pred = attrition_tree.predict(X_test)
from sklearn.metrics import confusion_matrix
confusionmatrix = confusion_matrix(Y_test, Y_pred)
print(confusionmatrix)
```

以上代码输出如下：

```
[[3289  139]
 [  92  980]]
```

看一下混淆矩阵，可以假设分类器对真阳率和真阴率的分类是可靠的。那根据总结出的评估指标对假设进行验证：

```
print('Accuracy of Decision Tree classifier on test set:
{:.2f}'.format(attrition_tree.score(X_test, Y_test)))
from sklearn.metrics import classification_report
print(classification_report(Y_test, Y_pred))
```

以上代码输出如下：

```
Accuracy of Decision Tree classifier on test set: 0.95
             precision    recall  f1-score   support

          0       0.97      0.96      0.97      3428
          1       0.88      0.91      0.89      1072

avg / total       0.95      0.95      0.95      4500
```

准确率和其他指标一样，是 0.95，很不错的分值。

决策树模型和结果比逻辑回归模型的结果更好。现在，我们再来了解另一种很常用的基于支持向量的分类建模法。

2.8 支持向量机
Support Vector Machines

SVM 是一种监督机器学习算法，主要用在分类任务中，但是，也可用来解决回归问题。

什么是 SVM
What is SVM?

SVM 是基于分割超平面原则的分类器。对已知训练数据集，SVM 算法会找出将分类间隔最大化的超平面，并利用这些间隔对新数据集做出预测。超平面是一个比其周围平面少一维度的子空间。二维数据集中，直线就是超平面。

在哪里使用 SVM
Where is SVM used?

SVM 与其他分类器的用例相似，但 SVM 更适合特征或属性数量比数据点或记录的数量大很多的场景。

使用什么方法实现 SVM
By Which Method Can SVM be Implemented?

构建 SVM 的过程与此前学到的分类方法类似。唯一的区别是从 scikit-learn 库中导入 SVM 方法。用 pandas 库导入人力流失数据集，分成 train 子集和 test 子集：

```
import numpy as np
import pandas as pd
from sklearn import svm
from sklearn.metrics import accuracy_score

hr_data = pd.read_csv('data/hr.csv', header=0)
hr_data.head()
hr_data = hr_data.dropna()
print(" Data Set Shape ", hr_data.shape)
print(list(hr_data.columns))
print(" Sample Data ", hr_data.head())
data_trnsf = pd.get_dummies(hr_data, columns =['salary', 'sales'])
data_trnsf.columns
X = data_trnsf.drop('left', axis=1)
X.columns
from sklearn.model_selection import train_test_split

X_train, X_test, Y_train, Y_test = train_test_split(X, data_trnsf.left,
test_size=0.3, random_state=42)
print(X_train)
```

接下来，创建一个 SVM 模型实例。我们将 kernel 设置为线性，因为要想一条直线分割两个类。很容易就可为线性可分割的数据找到最优超平面。然而，若数据不是线性可分的，则要把数据映射到一个新空间中，使其可以线性分割。这种方法称为 kernel 技巧（kernel trick）：

```
attrition_svm = svm.SVC(kernel='linear')
attrition_svm.fit(X_train, Y_train)
```

以上代码输出如下：

```
Out[9]: SVC(C=1, cache_size=200, class_weight=None, coef0=0.0,
        decision_function_shape=None, degree=3, gamma=1, kernel='linear',
        max_iter=-1, probability=False, random_state=None, shrinking=True,
        tol=0.001, verbose=False)
```

使用 SVM 模型实例拟合训练数据后，可对 test 数据集预测 Y 值，然后创建混淆矩阵，评估模型性能：

```
Y_pred = attrition_svm.predict(X_test)
from sklearn.metrics import confusion_matrix
confusionmatrix = confusion_matrix(Y_test, Y_pred)
print(confusionmatrix)
```

以上代码输出如下：

```
[[3212  216]
 [ 811  261]]
```

然后，计算出模型准确率和其他指标值：

```
print('Accuracy of SVM classifier on test set: {:.2f}'.format(attrition_svm.score(X_test, Y_test)))
from sklearn.metrics import classification_report
print(classification_report(Y_test, Y_pred))
```

上述代码输出如下：

```
Accuracy of SVM classifier on test set: 0.77
             precision    recall  f1-score   support

          0       0.80      0.94      0.86      3428
          1       0.55      0.24      0.34      1072

avg / total       0.74      0.77      0.74      4500
```

由此可见，带默认参数的 SVM 模型效果没有决策树的好。因此，到目前为止，决策树在人力流失预测积分榜上稳居榜首。那再试试另一种分类算法——K 近邻（k-nearest neighbors，KNN）算法，这是一种更易理解和使用但资源消耗却高得多的算法。

2.9 K 近邻算法
k-Nearest Neighbors

在为人力流失数据集构建 KNN 模型之前，先了解一下 KNN。

什么是 KNN
What is k-Nearest Neighbors?

KNN 是一种最直接的算法，能保存全部可用的数据点，根据距离相似性指标（如欧氏距离）预测新数据。这个算法可直接使用训练数据集做出预测。但这是一个高度资源密集型的算法，因为没有任何训练阶段，要求全部数据都放在内存中进行新实例预测。

 欧氏距离由两个点的方差和的平方根算出。

$$EuclideanDistance(x_i, y_i) = \sqrt{sum((x_i - y_i)^2)}$$

在哪里使用 KNN
Where is KNN used?

KNN 既可构建分类模型，也可构建回归模型。它可应用于二分和多分类的分类任务中。KNN 甚至可用来创建推荐系统或填充缺失值。使用方便，训练简单，结果也好解释。

使用什么方法实现 KNN
By Which Method Can KNN be Implemented?

同样，KNN 的构建也与前面模型的构建相似。从 scikit-learn 库中导入

KNeighborsClassifier 方法，使用 KNN 算法建模。然后，使用 pandas 库导入人力流失数据集，分成 train 子集和 test 子集：

```
import numpy as np
import pandas as pd
from sklearn.metrics import accuracy_score
from sklearn.neighbors import KNeighborsClassifier
hr_data = pd.read_csv('data/hr.csv', header=0)
hr_data.head()
hr_data = hr_data.dropna()
print(" Data Set Shape ", hr_data.shape)
print(list(hr_data.columns))
print(" Sample Data ", hr_data.head())
data_trnsf = pd.get_dummies(hr_data, columns =['salary', 'sales'])
data_trnsf.columns
X = data_trnsf.drop('left', axis=1)
X.columns
from sklearn.model_selection import train_test_split
X_train, X_test, Y_train, Y_test = train_test_split(X, data_trnsf.left,
test_size=0.3, random_state=42)
print(X_train)
```

创建 KNN 模型，需指明距离计算中要考虑的近邻数量。

 实践中建模时，会为一系列包含不同距离度量值的 n_neighbors 值构建不同的模型，然后选择准确率最高的那个模型。这个过程也就是**超参数优化**（tuning the hyperparameters）。

对于以下的人力流失模型，定义 n_neighbors 为 6，且距离指标为欧氏距离（Euclidean）：

```
n_neighbors = 6
attrition_knn = KNeighborsClassifier(n_neighbors=n_neighbors,
metric='euclidean')
attrition_knn.fit(X_train, Y_train)
```

以上代码输出如下：

```
Out[9]: KNeighborsClassifier(algorithm='auto', leaf_size=30, metric='euclidean',
            metric_params=None, n_jobs=1, n_neighbors=6, p=2,
            weights='uniform')
```

然后在 test 数据集中做预测，查看混淆矩阵和其他评估指标：

```
Y_pred = attrition_knn.predict(X_test)
from sklearn.metrics import confusion_matrix
confusionmatrix = confusion_matrix(Y_test, Y_pred)
print(confusionmatrix)
```

以上代码输出如下：

```
[[3273  155]
 [ 109  963]]
```

以下代码输出准确率和其他指标值：

```
print('Accuracy of KNN classifier on test set:
{:.2f}'.format(attrition_knn.score(X_test, Y_test)))
from sklearn.metrics import classification_report
print(classification_report(Y_test, Y_pred))
```

以上代码输出如下：

```
Accuracy of Decision Tree classifier on test set: 0.94
             precision    recall  f1-score   support

          0       0.97      0.95      0.96      3428
          1       0.86      0.90      0.88      1072

avg / total       0.94      0.94      0.94      4500
```

KNN 的结果比 SVM 模型的结果好一些，但仍比决策树的得分低。KNN 是一种资源密集型算法。如果使用 KNN 只有小幅改善，那么仍推荐使用其他算法模型。但是，这最终取决于用户的考量，依据环境条件和目标问题选择最合适的算法。

2.10 集成方法
Ensemble Methods

集成模型是一种提高预测模型有效性的健壮方法。这是一种很周全的策略，跟一个说出来就充满力量的词很相似——团队！一个任务由团队协作完成就能取得很显著的成就。

集成模型是什么
What are Ensemble Models?

在机器学习的世界也一样，集成模型是一组模型，一起工作，一起提升工作效率。从技术上讲，集成模型由几种独立训练的监督学习模型组成，通过不同的方式合并结果，实现最终的预测。相比组合中任意一个学习算法单独的预测，这种预测结果更为准确。

大多数情况下，有以下 3 种集成学习方法。

- 套袋法（bagging）。
- 提升（boosting）。
- 叠加/融合（stacking/blending）。

套袋法

套袋法（bagging）也称**自举汇聚**（bootstrap aggregation）。这是一种减小模型结果方差误差的方法。有一些脆弱的学习算法很敏感——稍有不同的输入就可以产生完全离谱的输出。随机森林通过运行多个实例来降低这种可变性，减小方差。套袋法中，从训练数据集中用各种替代模型提取随机样本的方法反复抽取随机样本（引导过程）。

模型是在每个样本上使用监督学习训练出来的。最后，通过取预测均值或多数投票技术，选出最佳预测，并将结果进行合并。多数投票是指在全部分类器中选取预测数量最多的类作为总集成预测。还有很多其他方法可用于确定最终结果，比如权重和排名平均法。

scikit-learn 中有各种各样的套袋算法，如套袋决策树、随机森林、其他树等。下面会演示最流行的随机森林模型，其他的模型你可以自行尝试。从 scikit-learn 的 `ensemble` 包中导入 `RandomForestClassifier`，实现随机森林。还是使用人力流失数据，这次演示的部分代码也与前面的相同：

```
import numpy as np
import pandas as pd
from sklearn.ensemble import RandomForestClassifier
from sklearn.metrics import accuracy_score
from sklearn import tree
hr_data = pd.read_csv('data/hr.csv', header=0)
hr_data.head()
hr_data = hr_data.dropna()
print(" Data Set Shape ", hr_data.shape)
print(list(hr_data.columns))
print(" Sample Data ", hr_data.head())
data_trnsf = pd.get_dummies(hr_data, columns =['salary', 'sales'])
```

```
data_trnsf.columns
X = data_trnsf.drop('left', axis=1)
X.columns
from sklearn.model_selection import train_test_split
X_train, X_test, Y_train, Y_test = train_test_split(X, data_trnsf.left,
test_size=0.3, random_state=42)
print(X_train)
```

随机森林模型实例化没有强制性参数要求。但是，要构建良好的随机森林模型，还是要了解一些参数的，描述如下。

- `n_estimators`：指明在模型中要创建的树的数量。默认为 10。
- `max_features`：指明每个分隔中随机选为候选变量或特征的数量。默认为 $\sqrt{number\_of\_features}$

设 n_estimators 为 100，max_features 为 3，构建随机森林模型，代码如下：

```
num_trees = 100
max_features = 3
attrition_forest = RandomForestClassifier(n_estimators=num_trees,
max_features=max_features)
attrition_forest.fit(X_train, Y_train)
```

以上代码输出如下：

```
Out[6]: RandomForestClassifier(bootstrap=True, class_weight=None, criterion='gini',
            max_depth=None, max_features=3, max_leaf_nodes=None,
            min_impurity_split=1e-07, min_samples_leaf=1,
            min_samples_split=2, min_weight_fraction_leaf=0.0,
            n_estimators=100, n_jobs=1, oob_score=False, random_state=None,
            verbose=0, warm_start=False)
```

模型拟合成功后，在 test 或预留数据集中预测 Y_pred 值：

```
Y_pred = attrition_forest.predict(X_test)
from sklearn.metrics import confusion_matrix
confusionmatrix = confusion_matrix(Y_test, Y_pred)
print(confusionmatrix)
```

混淆矩阵中的结果看起很不错，分类错误更少，预测准确。看一下评估指标如何：

```
[[3417   11]
 [  52 1020]]
```

下一步查看 Random Forest Classifier 的准确率以及 print 分类报告：

```
print('Accuracy of Random Forest classifier on test set:
```

```
{:.2f}'.format(attrition_forest.score(X_test, Y_test)))
from sklearn.metrics import classification_report
print(classification_report(Y_test, Y_pred))
```

以上代码输出如下:

```
Accuracy of Random Forest classifier on test set: 0.99
             precision    recall  f1-score   support

          0       0.99      1.00      0.99      3428
          1       0.99      0.95      0.97      1072

avg / total       0.99      0.99      0.99      4500
```

模型很出色，全部评估指标接近完美预测。结果好到不敢相信，有可能会过度拟合。然而，暂时先把随机森看成是人力流失数据集的最佳算法，继续看另一种广泛使用的组合模型技术——提升。

提升

提升（boosting）是一个迭代的过程，新模型是在上一个模型的缺陷基础上构建起来的，一个一个交替构建。有助于减少模型偏差，也会减小方差。提升法会在上一模型表现较差的基础上，改善生成新分类器，以便更好地预测。与套袋法不同，训练数据的重新取样取决于上一个分类器的表现。提升法使用全部数据训练单个分类器，但之前被分类器分错的实例受到重点关注，以便于后续分类器提升结果。

梯度提升机（gradient boosting machines，GBMs），也称**随机梯度提升**（stochastic gradient boosting，SGB），是提升法的一个示例。再次导入所需软件包，加载人力流失数据集。同样，也执行相同的流程，把类别数据集转换成独热编码值，把数据集按 70:30 比例分割成 train 子集和 test 子集:

```
import numpy as np
import pandas as pd
from sklearn.ensemble import GradientBoostingClassifier
from sklearn.metrics import accuracy_score
from sklearn import tree
hr_data = pd.read_csv('data/hr.csv', header=0)
hr_data.head()
hr_data = hr_data.dropna()
print(" Data Set Shape ", hr_data.shape)
print(list(hr_data.columns))
```

```
print(" Sample Data ", hr_data.head())

data_trnsf = pd.get_dummies(hr_data, columns =['salary', 'sales'])
data_trnsf.columns
X = data_trnsf.drop('left', axis=1)
X.columns

from sklearn.model_selection import train_test_split
X_train, X_test, Y_train, Y_test = train_test_split(X, data_trnsf.left,
test_size=0.3, random_state=42)
print(X_train)
```

有些很重要的 GradientBoostedClassifier 的最优参数,但不全是必选的。

- n_estimators:与随机森林算法的 n_estimators 相似,但这里的树是有序创建的,在提升法的不同阶段。使用这些参数指明模型中树的数量或提升阶段的数量。默认值为 100。

- max_features:这是找到最佳分割时需考虑的特征数量。max_features 比特征数量少的时候,方差会减小,但模型的偏差会增高。

- max_depth:每棵树的最大树深。默认值为 3:

```
num_trees = 100
attrition_gradientboost=
GradientBoostingClassifier(n_estimators=num_trees, random_state=42)
attrition_gradientboost.fit(X_train, Y_train)
```

以上代码输出如下:

```
Out[6]: GradientBoostingClassifier(criterion='friedman_mse', init=None,
                learning_rate=0.1, loss='deviance', max_depth=3,
                max_features=None, max_leaf_nodes=None,
                min_impurity_split=1e-07, min_samples_leaf=1,
                min_samples_split=2, min_weight_fraction_leaf=0.0,
                n_estimators=100, presort='auto', random_state=42,
                subsample=1.0, verbose=0, warm_start=False)
```

模型成功拟合数据集后,使用训练过的模型预测 test 子集的 Y 值:

```
Y_pred = attrition_gradientboost.predict(X_test)

from sklearn.metrics import confusion_matrix
confusionmatrix = confusion_matrix(Y_test, Y_pred)
print(confusionmatrix)
```

以下混淆矩阵看起来不错,分类错误误差最小:

```
[[3389   39]
 [  85  987]]
```

打印准确率和其他指标来评估分类器：

```
print('Accuracy of Gradient Boosting Classifier classifier on test set:
{:.2f}'.format(attrition_gradientboost.score(X_test, Y_test)))
from sklearn.metrics import classification_report
print(classification_report(Y_test, Y_pred))
```

以上代码输出如下：

```
Accuracy of Gradient Boosting Classifier classifier on test set: 0.97
             precision    recall  f1-score   support

          0       0.98      0.99      0.98      3428
          1       0.96      0.92      0.94      1072

avg / total       0.97      0.97      0.97      4500
```

准确率为 97%，很不错，但没有随机森林模型的高。后面的章节再讨论另一种集成模型。

叠加/融合

本方法中会有多层的分类器叠加/融合在一起。第一层分类器的预测概率用于训练第二层分类器，以此类推。最终结果是通过如逻辑回归之类的一个基础分类器实现的。也可使用不同的算法，如决策树、随机森林或 GBM 作为最后一层分类器。

在 scikit-learn 中没有现成的叠加集成的实现可用。但在第 4 章中，我们会演示使用 scikit-learn 的基础算法创建自动化叠加集成函数。

2.11 分类器结果对比
Comparing the Results of Classifiers

我们已经基于人力流失数据集创建了约 6 种分类模型。图 2-9 总结了每种模型的评分。

	准确率	精度	召回	F1得分
逻辑回归	0.79	0.76	0.79	0.76
决策树	0.95	0.95	0.95	0.95
SVM	0.77	0.74	0.77	0.74
KNN	0.94	0.94	0.94	0.94
随机森林	0.99	0.99	0.99	0.99
梯度提升分类器	0.97	0.97	0.97	0.97

图 2-9 分类模型评分

从图 2-9 可以看出，随机森林以 99% 的破纪录准确率在全部 6 种模型中胜出。现在，随机森林模型无须再改进，但要验证它是否对新数据集也能良好泛化，以防结果与 train 数据集过度拟合。方法之一就是进行交叉验证。

2.12 交叉验证
Cross-Validation

交叉验证（cross-validation）是评估一个模型在非训练数据集上预测准确率的方法，非训练数据集是指训练模型未知的数据样本。交叉验证是要确保模型部署到生产环境中时在新的独立数据集上泛化良好。一种方法是将数据集分为两个子集：训练子集和测试子集。前面的示例中已经演示过这种方法。

另一种更受欢迎且更可靠的方法是 k-折交叉验证法，即将一个数据集划分为 k 个相同大小的子样本。其中 k 是非 0 正整数。在训练阶段，会将 $k-1$ 个样本用于训练模型，剩下的一个样本用于测试模型。这个过程会重复 k 次，其中 k 个样本中的每一个只用过一次测试模型。然后将评估结果求平均，或者通过多数投票等方法综合评估，最后提供一个估值。

我们对前面构建的随机森林模型进行 5 和 10 折交叉验证，从而评估其表现。在随机森林代码后，添加以下代码片段：

```
crossval_5_scores = cross_val_score(attrition_forest, X_train, Y_train, cv=5)
print(crossval_5_scores)
print(np.mean(crossval_5_scores))
crossval_10_scores = cross_val_score(attrition_forest, X_train, Y_train, cv=10)
print(crossval_10_scores)
print(np.mean(crossval_10_scores))
```

5-折和 10-折交叉验证的准确率得分分别为 0.9871 和 0.9875。得分很高，与实际的随机森林模型 0.99 分非常接近，如下截图。这就保证了此模型可很好地泛化到其他独立的数据集。

```
[0.98714286 0.98571429 0.98571429 0.98809524 0.9890424 ]
0.9871418135620136
[0.98857143 0.98666667 0.98666667 0.98571429 0.98285714 0.98857143
 0.98857143 0.98857143 0.98952381 0.98951382]
0.9875228108402562
```

对监督机器学习有了基本理解之后，则继续进入无监督机器学习。

在本章开始部分介绍过无监督机器学习。重温下目标。

无监督学习的目标是通过推理输入数据集属性的结构和关系而识别出其中的模式。

那么，识别模式有什么算法和方法可用呢？有很多，如聚类和自编码器。下面会介绍聚类，并且在第 7 章中讲解自编码器。

2.13　聚类
Clustering

本节我们来问一个问题。如何开始学习一个新算法或一种机器学习方法？先来说说 3W。那就从聚类的 3W 开始吧。

什么是聚类
What is Clustering?

聚类（clustering）是一种将相似数据分组的技术，每组都有与其他组不同的特性。数据聚类有多种方法。其中一种方法是基于规则的，依据预定义条件进行分组，如根据年龄或行业给客户分组。另一种方法是使用机器学习算法将数据聚类到一起。

在哪里使用聚类
Where is Clustering Used?

无监督学习过程最常用于从行业数据中推演出逻辑关系和模式。聚类可跨行业应用，也可用于商业领域。利用聚类可进行信息提取、客户分类、图像分类，以及网页、新闻等非结构化文本的聚类。

用什么方法实现聚类
By Which Method Can Clustering be Implemented?

创建聚类有各种机器学习方法。聚类算法属于以下分类之一。

- **层次聚类**（hierarchical clustering）：也称**凝聚聚类**（aggolmerative clustering），是把每个数据点与其距离上最近的点连接起来。这是一个递归的过程，从一条记录开始，配对，不断迭代，直到全部记录组成一个聚类。想象一下，它的结构与倒立的树是相似的，可视觉化为一个系统树图。采用这种方法的问题之一是确定聚类的过程。这个过程非常消耗资源，但可看到系统树状图，选择聚类的数量。

- **划分聚类**（partition-based clustering）：这种方法中，数据被分割成不同的分区。分区是依据数据点之间的距离进行的。k-means 算法是一种常用的划分聚类法。其中，不同的距离函数会影响聚类的形状。欧氏（Euclidean）距离、曼哈顿（Manhattan）距离和余弦（cosine）距离是创建 k-means 聚类时用得最多的三种距离函数。欧氏距离对输入向量的大小是最敏感的。这时就要对输入向量的数值范围进行归一化或标准化处理，或选择一种对大小不敏感的距离度量法，如余弦距离。

- **密度技术**（density-based technique）：这种技术是通过使用数据点特定概率分布而形成聚类的。主要理念是只要周边环境的密度超过定义阈值，就要继续扩展聚类。高密度区域会被标记为聚类，与低密度区域隔离，低密度区域可能为干扰。干扰是随机变异或数据误差，是统计学上不确定的或无法解释的。

- **网格法**（grid-based method）：本方法中，首先是分割数据集属性，创建超矩形网格单元。然后，将低于所定义的阈值参数的低密度单元丢弃。那周围的高密度单元就会聚集在一起，直到目标函数实现或保持恒定为止。最终得出的风格单元就被视为聚类。

接下来详细了解另外两种业务广泛使用的方法：层次聚类和 k-means 聚类。

层次聚类
Hierarchical Clustering

我们在 Python 中使用 scikit-learn 进行层次聚类。从 sklearn.cluster 中导入 `AgglomerativeClusteing` 方法，创建聚类。层次聚类适用于距离度量指标，因此要在建模前把类别型数据转换成合适的数值型格式。前面用独热编码做过类别属性到数值格式的转换，其实还有很多方法可以用。第 3 章会详细描述，代码如下：

```
import pandas as pd
import numpy as np
from sklearn import preprocessing
from sklearn.cluster import AgglomerativeClustering
hr_data = pd.read_csv('data/hr.csv', header=0)
hr_data.head()
hr_data = hr_data.dropna()
print(hr_data.shape)
print(list(hr_data.columns))
data_trnsf = pd.get_dummies(hr_data, columns =['salary', 'sales'])
data_trnsf.columns
```

接下来使用下面的参数对 `AgglomeartiveClustering` 进行实例化，使用模型拟合数据。

- `n_clusters`：要发现的聚类的数量。默认值为 2。
- `affinity`：用于计算连接的距离指标。默认为 `euclidean`，也可以使用 `mahattan`、`cosine`、`L1`、`L2` 和 `precomputed` 等其他距离指标。
- `linkage`：此参数决定成对合并聚类所用到的指标。不同的连接指标有：
    - Ward：将要合并的聚类方差降至最小。这是默认参数值。
    - Average：采用两个集的每个观测值间距的平均值。
    - Complete：采用两个集的全部观测值中的最大值。

现在使用一些上述参数构建 `AgglomerativeClustering` 模型：

```
n_clusters = 3
clustering = AgglomerativeClustering(n_clusters=n_clusters,
affinity='euclidean', linkage='complete')
clustering.fit(data_trnsf)
cluster_labels = clustering.fit_predict(data_trnsf)
```

模型建好后，要对其进行评估。评估聚类结果的最佳方式是人工检查所形成的聚类，确定每个聚类代表什么，每个聚类中的数据有什么共同的值。

与人工检查相结合，也可采用轮廓系数（silhouette_score）来确定最佳模型。轮廓系数范围在−1和+1之间：

- +1 表示聚类数据与指定聚类相接近，与相邻聚类距离远；
- −1 表示数据点与相邻聚类更近，与指定聚类更远。

当模型的平均轮廓系数为−1时，说明模型很差，而得+1分的则说明模型很理想。所以，平均轮廓系数越高，聚类模型越好：

```
silhouette_avg = silhouette_score(data_trnsf, cluster_labels)
print("For n_clusters =", n_clusters,"The average silhouette_score is :",
silhouette_avg)
```

上述代码输出如下：

```
For n_clusters = 3 The average silhouette_score is : 0.496869735878
```

此模型的平均轮廓系数为 0.49，可以初步判断所形成的聚类不错。

比较这个分数和 k-means 聚类结果，选出最好的模型，在人力流失数据集上创建三个聚类。

划分聚类（KMeans）
Partitioning Clustering (KMeans)

从 scikit-learn 包中导入 KMeans 法，其余代码与层次聚类相似：

```
import pandas as pd
import numpy as np
from sklearn import preprocessing
import matplotlib.pyplot as plt
from sklearn.cluster import KMeans
from sklearn.metrics import silhouette_samples, silhouette_score
hr_data = pd.read_csv('data/hr.csv', header=0)
hr_data.head()
```

```
hr_data = hr_data.dropna()
print(hr_data.shape)
print(list(hr_data.columns))
data_trnsf = pd.get_dummies(hr_data, columns =['salary', 'sales'])
data_trnsf.columns
```

建模前，要指明 k-means 函数中的聚类数量（n_clusters）。这是创建 k-means 聚类的必要参数。默认值为 8。下一步将数据拟合到 KMeans 实例中，然后创建模型。fit_predict 运算这些值，获取聚类标签，与在 AggloerativeClustering 的步骤相同：

```
n_clusters = 3
kmeans = KMeans(n_clusters=n_clusters)
kmeans.fit(data_trnsf)
cluster_labels = kmeans.fit_predict(data_trnsf)
```

可采用 cluster_centers_ 和 means_labels_ 查看聚类形心和标签：

```
centroid = kmeans.cluster_centers_
labels = kmeans.labels_
print (centroid)
print(labels)
silhouette_avg = silhouette_score(data_trnsf, cluster_labels)
print("For n_clusters =", n_clusters,"The average silhouette_score is :", silhouette_avg)
```

以上代码输出如下：

```
[[  6.75324232e-01   7.28759954e-01   3.82161547e+00   2.02696701e+02
    3.41501706e+00   1.71786121e-01   6.05233220e-02   2.38907850e-02
    9.67007964e-02   4.67121729e-01   4.36177474e-01   7.73606371e-02
    5.09670080e-02   4.73265074e-02   4.73265074e-02   5.05119454e-02
    5.62002275e-02   6.07508532e-02   2.73037543e-01   1.51535836e-01
    1.84982935e-01]
 [  5.81896364e-01   6.47789091e-01   3.25945455e+00   1.46582364e+02
    3.31472727e+00   1.34545455e-01   2.93090909e-01   2.03636364e-02
    7.74545455e-02   5.00909091e-01   4.21636364e-01   7.96363636e-02
    5.18181818e-02   5.47272727e-02   5.23636364e-02   3.83636364e-02
    5.94545455e-02   6.12727273e-02   2.79454545e-01   1.48909091e-01
    1.74000000e-01]
 [  5.92360893e-01   7.78814655e-01   4.37284483e+00   2.58326607e+02
    3.76763323e+00   1.32053292e-01   3.31700627e-01   1.99843260e-02
    7.56269592e-02   4.91379310e-01   4.32993730e-01   8.79702194e-02
    5.44670846e-02   5.05485893e-02   4.76097179e-02   3.85971787e-02
    5.56426332e-02   5.83855799e-02   2.74882445e-01   1.45768025e-01
    1.86128527e-01]]
[1 2 2 ..., 1 2 1]
For n_clusters = 3 The average silhouette_score is : 0.581903917045
```

k-means 聚类的平均 silhouette_score 为 0.58，比层次聚类的平均轮廓系数更高。

说明人力流失数据集在 k-means 模型下形成的三个聚类比在层次模型中形成的聚类更好。

2.14 总结
Summary

机器学习及其自动化的过程是漫长的。本章的目的是熟悉机器学习概念；更重要的是，熟悉 scikit-learn 和其他 Python 包，这样就可以顺利地加快后面几章的学习，我们创建线性回归模型和 6 种分类模型，并且了解聚类技术，以及对比各个模型。

本章全部的分类器都是在同一个人力流失数据集基础上创建起来的。我们发现这些代码之间有很多相似之处，导入的库也非常相似，只有一个机器学习分类实例化的库不同。数据预处理模块在所有代码中都是冗余的。机器学习技术根据任务和目标属性数据而变化。同样，评估方法也与机器学习方法对等。

是不是觉得有些部分冗余而应该自动化？对，可以自动化，但却没那么简单。初始考虑自动化时，要把模型本身和周边因素都结合起来。每个代码片段都是一个独立的模块。

第 3 章内容完全是关于数据预处理模块的。这是机器学习项目中最重要、也最耗时的主题。我们会讨论创建一个可靠机器学习模型所必需的各种数据准备任务。

第 3 章
数据预处理
Data Preprocessing

但凡对**机器学习**感兴趣的人，应该都听说过数据科学家或机器学习工程师80%的时间都在准备数据，其余20%的时间在进行模型构建和评估。准备过程中耗费的大量时间算作是构建良好模型的一种投资吧。在好的数据集上训练出的简单模型也比在差的数据集上跑出来的复杂模型要强。实践中，很难找到靠谱的数据集。高质量的数据集是靠自己创建和整理出来的。怎么构建好的数据呢？这就是本章要讨论的内容之一。从理论上讲，好不好是相对于手头的任务以及如何处理和消耗数据而言的。本章会覆盖以下主题。

- 数据转换（data transformation）。
- 特征选择（feature selection）。
- 降维（dimensionality reduction）。
- 特征生成（feature generation）。

我们将逐一讨论这些处理方式，并介绍用于特征准备的常用 Python 开源工具。

下面先从数据转换开始说起。

3.1 技术要求
Technical Requirements

全部代码示例均放在 GitHub 的 `Chapter 03` 文件夹中。

3.2 数据转换
Data Transformation

假设要构建一个机器学习模型，任务是预测员工流失。在理解业务的基础上可以加一些相关变量，有助于创建好模型。另外，也可以丢弃一些没有价值的特征，如 `EmployeeID`。

识别 ID 列的过程也称标识检测（identifier detection）。ID 列在模式检测和预测中对模型没有什么价值。因此，ID 列检测可算作是 `AutoML` 包的一个功能，依据具体算法或任务是否依赖 ID 决定是否使用这一功能。

确定好字段范围后，就可以探索数据，将可能对学习过程有用的特征进行转换。数据转换的过程也会给数据增加可用价值，有利于机器学习模型的构建。比如，员工入职日期 11-02-2018 是没有信息量的。但是，如果把这个特征转换成 4 个属性——日期、日、月和年，就对模型的构建提供了价值。

特征转换也与所用的机器学习算法类型是强关联的。广义上，监督模型可分为两个类别——树模型和非树模型。

树模型可自行处理大多数特征中的异常值。非树模型，如近邻算法和线性回归，要先进行特征转换，模型的预测能力才会提升。

原理解释不再赘述。现在直奔主题，看一些常见的特征转换方法，适用于各种数据类型。先从数值型特征开始。

数值型数据的转换
Numerical Data Transformation

以下是数值型数据转换最常用的方法。

- 缩放（scaling）。
- 缺失值（missing values）。
- 异常值（outliers）。

以上技术可嵌入函数中，直接应用在机器学习流水线中，进行数值型数据的转换。

缩放

标准化（standardization）和**归一化**（normalization）是业内**缩放**（scaling）技术的两个术语。两种技术都是为了保证模型中的数值型特征及其代表的权重等价。很多时候，标准化和归一化可互换使用。虽然两种都属于缩放技术，但还是有一丝细微差别。

标准化保证数据正态分布。标准化会调整数据使其均值为 0 且方差为 1。归一化是假设数据无先验分布。把数值型数据重新调整到固定的范围中：0 到 1，-1 到+1 等。

以下是一些常用的数据标准化和归一化技术。

- **Z 分标准化**（Z-score standardization）：如果数据符合高斯分布，则缩放数据到均值为 0 且标准方差为 1。前提条件之一是数值型数据呈正态分布。数学上，符号表示为 $Z = X - \bar{X}/\sigma$，其中 $\bar{X}$ 是平均值，σ 为标准方差。

scikit-learn 提供各种数据标准化和归一化方法。首先，使用以下代码加载人力流失数据集：

```
%matplotlib inline
import numpy as np
import pandas as pd
hr_data = pd.read_csv('data/hr.csv', header=0)
print (hr_data.head())
```

上述代码输出了数据集的各种属性以及一些数据点。

```
        satisfaction_level  last_evaluation  number_project  average_montly_hours  \
0                 0.38             0.53               2                   157
1                 0.80             0.86               5                   262
2                 0.11             0.88               7                   272
3                 0.72             0.87               5                   223
4                 0.37             0.52               2                   159

   time_spend_company  Work_accident  left  promotion_last_5years  sales  \
0                   3              0     1                      0  sales
1                   6              0     1                      0  sales
2                   4              0     1                      0  sales
3                   5              0     1                      0  sales
4                   3              0     1                      0  sales

   salary
0     low
1  medium
2  medium
3     low
4     low
```

使用以下代码分析数据集的分布：

hr_data[hr_data.dtypes[(hr_data.dtypes=="float64")|(hr_data.dtypes=="int64")].index.values].hist(figsize=[11,11])

以上代码输出了一些不同数值型属性的直方图，如图 3-1 所示。

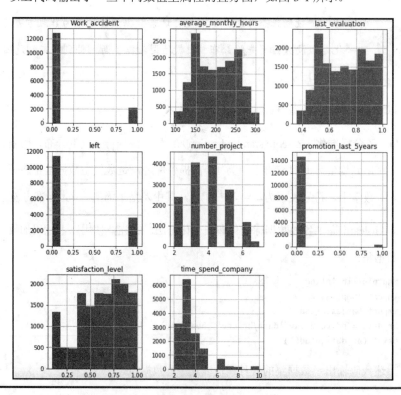

图 3-1　属性直方图

作为示例，我们可以使用 `sklearn.preprocessing` 模块中的 `StandardScaler` 对 satisfaction_level 列的值进行标准化。导入方法后，首先要创建 `StandardScaler` 类的一个实例。接着使用 `fit_transform` 方法对该列进行拟合和转换。下面的示例中使用的是待标准化的 satisfaction_level 属性：

```
from sklearn.preprocessing import StandardScaler
scaler = StandardScaler()
hr_data_scaler=scaler.fit_transform(hr_data[['satisfaction_level']])
hr_data_scaler_df = pd.DataFrame(hr_data_scaler)
hr_data_scaler_df.max()
hr_data_scaler_df[hr_data_scaler_df.dtypes[(hr_data_scaler_df.dtypes=="float64")|(hr_data_scaler_df.dtypes=="int64")].index.values].hist(figsize=[11,11])
```

执行以上代码后，再次查看 satisfaction_level 实例直方图，如图 3-2 所示，可观察到这些数值已经标准化到−2 到 1.5 之间了。

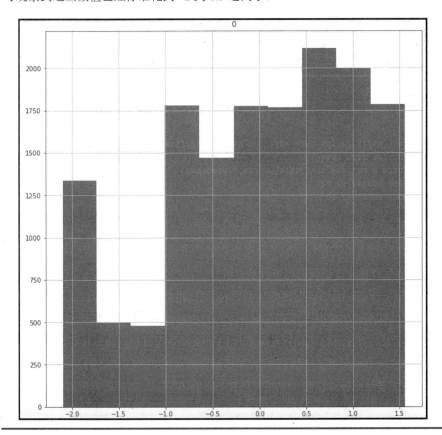

图 3-2　satisfaction_level 实例直方图

- **最小-最大归一化**（min-maxnormalization）：从变量的实际值中减去其最小值，除以其最大值与最小值的差。数学上表示如下：

$$z = \frac{x - min(x)}{max(x) - min(x)}$$

scikit-learn 的 preprocessing 模块中有 MinMaxScaler 方法。本例中，将人力流失数据集的 4 个属性归一化，即 average_monthly_hours（平均每月小时数）、last_evaluation（最近评估）、number_project（项目数量）以及 satisfaction_level（满意度）。步骤与 StandardScaler 中的相似。首先从 sklearn.preprocessing 模块中导入 MinMaxScaler，创建 MinMaxScaler 类的实例。

接着，使用 fit_transform 方法对属性列进行拟合和转换：

```
from sklearn.preprocessing import MinMaxScaler
minmax=MinMaxScaler()
hr_data_minmax=minmax.fit_transform(hr_data[[ 'average_montly_hours',
 'last_evaluation', 'number_project', 'satisfaction_level']])
hr_data_minmax_df = pd.DataFrame(hr_data_minmax)
hr_data_minmax_df.min()
hr_data_minmax_df.max()
hr_data_minmax_df[hr_data_minmax_df.dtypes[(hr_data_minmax_df.dtypes=="float64")|(hr_data_minmax_df.dtypes=="int64")].index.values].hist(figsize=[11,11])
```

如图 3-3 所示，转换后 4 个属性值分布在 0 和 1 之间。

缺失值

我们常碰到一些数据集里有一些变量或属性没有值。这可能有几种原因，比如调研的问题被忽略了、输入失误、设备故障，等等。在数据挖掘项目中遇到缺失值是正常的，而且有必要进行处理。

数据科学家被迫花费大量时间填充缺失值。填充缺失值的方法有许多种。用什么填充成了决定性因素。决定使用什么填充和什么时候填充是一种技巧，源于数据处理过程中积累的经验。有时，最好的办法是直接丢弃，但有些任务中最好把缺失值补充完整。

这时，就出现两个重要的问题。

- 什么时候使用哪种填充办法？
- 填充缺失值使用什么方法最好？

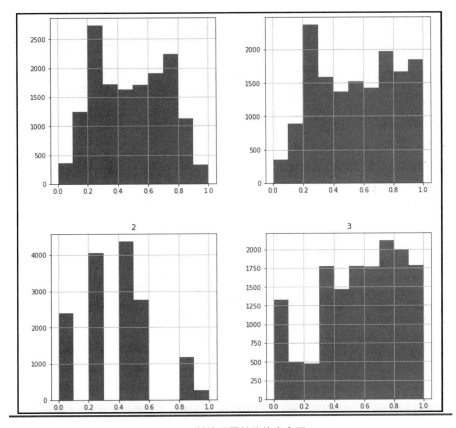

图 3-3 转换后属性值的直方图

我们认为这依赖于缺失值处理的经验。最好的办法是对数据应用不同的填充方法做一次对比研究，然后选出最合适的方法，给缺失值赋予偏差最小的值。

概括而言，遇到有缺失值时，首先尽量查出为什么缺失这个值。是收集数据时的问题导致的，还是因为数据源本身有问题。从根源上修复问题，比直接填充数值要更有用。比如，调查数据集中，有些调查对象不愿意透露某些信息，那就不得不补充所缺失的信息。

所以，填充缺失值之前，可采用以下准则。

- 调查缺失数据（investigate the missing data）；
- 分析缺失数据（analyze the missing data）；
- 确定生成最小偏差估值的最佳策略（decide the best strategy that yields least-biased estimates）。

可简化记为缺失值填充的 IAD 准则（即 investigate 调查、analyze 分析、decide 确定）。以下是一些填充缺失值的方法。

（1）**丢弃或删除数据**（remove or delete the data）：缺失的数据很少时，就可以忽略不计，单独分析，这种方法称为**成列删除**（list wise deletion）。但是，如果缺失值太多，就不建议采用这种方法，因为可能会错过数据中有价值的信息。**成对删除**（pairwise deletion）是另一种技术，只删除缺失的值。这是指我们只分析有值的用例。这是一种安全的策略，但这种方法的问题是，即使数据中每次只有很小的变动，从样本中得到的结果也会不同。

仍然使用人力流失数据集演示缺失值的处理。首先加载数据集，查看其中空值的数量：

```
import numpy as np
import pandas as pd
hr_data = pd.read_csv('data/hr.csv', header=0)
print (hr_data.head())
print('Nulls in the data set' ,hr_data.isnull().sum())
```

从以下的输出结果可看到，数据集相对干净，只有 promotion_last_5years 有一些缺失值。因此，我们给其中一些列制造缺失值：

```
   satisfaction_level  last_evaluation  number_project  average_montly_hours  \
0                0.38             0.53               2                   157
1                0.80             0.86               5                   262
2                0.11             0.88               7                   272
3                0.72             0.87               5                   223
4                0.37             0.52               2                   159

   time_spend_company  Work_accident  left  promotion_last_5years  sales  \
0                   3              0     1                      0  sales
1                   6              0     1                      0  sales
2                   4              0     1                      0  sales
3                   5              0     1                      0  sales
4                   3              0     1                      0  sales

   salary
0     low
1  medium
2  medium
3     low
4     low
Nulls in the data set satisfaction_level       0
last_evaluation          0
number_project           0
average_montly_hours     0
time_spend_company       0
Work_accident            0
left                     0
promotion_last_5years    0
sales                    0
salary                   0
dtype: int64
```

用以下代码把 promotion_last_5years、average_montly_hours 和 number_project 的一些值替换为空值:

```
#As there are no null introduce some nulls by replacing 0 in
promotion_last_5years with NaN
hr_data[['promotion_last_5years']] = hr_data[[
'promotion_last_5years']].replace(0, np.NaN)
#As there are no null introduce some nulls by replacing 262 in
promotion_last_5years with NaN
hr_data[['average_montly_hours']] = hr_data[[
'average_montly_hours']].replace(262, np.NaN)
#Replace 2 in number_project with NaN
hr_data[['number_project']] = hr_data[[
'number_project']].replace(2, np.NaN)
print('Nulls in the data set', hr_data.isnull().sum())
```

执行代码,这三列中插入了一些空值,结果如下所示:

```
Nulls in the data set satisfaction_level      0
last_evaluation          0
number_project        2388
average_montly_hours    86
time_spend_company       0
Work_accident            0
left                     0
promotion_last_5years 14680
sales                    0
salary                   0
dtype: int64
```

先复制一份 hr_data,不改变原始数据集,原始数据集还要用于演示其他缺失值填充方法。接下来,使用 dropna 方法丢弃有空值的行:

```
#Remove rows
hr_data_1 = hr_data.copy()
print('Shape of the data set before removing nulls ',
hr_data_1.shape)
# drop rows with missing values
hr_data_1.dropna(inplace=True)
# summarize the number of rows and columns in the dataset
print('Shape of the data set after removing nulls
',hr_data_1.shape)
```

如下可见,执行此操作后,行数从 14999 行减少到了 278 行。删除行的方法要谨慎使用。由于 promotion_last_5years 有 14680 个空值,所以会删除掉 14680 条记录:

```
Shape of the data set before removing nulls  (14999, 10)
Shape of the data set after removing nulls  (278, 10)
```

（2）**使用全局常量填充缺失值**（use a global constant to fill in the missing value）：可以使用全局常量，如 NA 或 –999，将缺失值与数据集的其他数据区分开。而且，虽然有些空值本身没有值，但却是数据集的重要组成部分。这些值是故意留空的。当无法区分空值与缺失值时，万全之策就是使用全局常量替代缺失值。

可以使用 fillna 方法把缺失值替换为 –999 之类的常量。以下代码演示这种方法的用法：

```
#Mark global constant for missing values
hr_data_3 = hr_data.copy()
# fill missing values with -999
hr_data_3.fillna(-999, inplace=True)
# count the number of NaN values in each column
print(hr_data_3.isnull().sum())
print(hr_data_3.head())
```

结果如下，从中可看出全部缺失值已经被替换为 –999：

```
satisfaction_level     0
last_evaluation        0
number_project         0
average_montly_hours   0
time_spend_company     0
Work_accident          0
left                   0
promotion_last_5years  0
sales                  0
salary                 0
dtype: int64
   satisfaction_level  last_evaluation  number_project  average_montly_hours  \
0                0.38             0.53          -999.0                 157.0
1                0.80             0.86             5.0                -999.0
2                0.11             0.88             7.0                 272.0
3                0.72             0.87             5.0                 223.0
4                0.37             0.52          -999.0                 159.0

   time_spend_company  Work_accident  left  promotion_last_5years  sales  \
0                   3              0     1                 -999.0  sales
1                   6              0     1                 -999.0  sales
2                   4              0     1                 -999.0  sales
3                   5              0     1                 -999.0  sales
4                   3              0     1                 -999.0  sales

   salary
0     low
1  medium
2  medium
3     low
4     low
```

（3）**使用属性均值/中值替换缺失值**（replace missing values with the attribute mean/median）：这种方法是数据科学家和机器学习工程师最可能用到的。数值型缺失值可以使用其均值或中值替换，类别型缺失值可以使用众数替换。这种方法的不足是可能会降低属性的可变性，继而弱化相关性的估算。如果处理监督分类模型，则可以使用组均值或组中值替换数值型缺失值，使用组众数替换类别型缺失值。组均值/中值替换方法中，按目标值对属性值分组，组内缺失的值就使用组均值/中值代替。

这里还可以使用 fillna 方法，应用平均函数，用均值替换缺失值。以下代码展示了其用法：

```
#Replace mean for missing values
hr_data_2 = hr_data.copy()
# fill missing values with mean column values
hr_data_2.fillna(hr_data_2.mean(), inplace=True)
# count the number of NaN values in each column
print(hr_data_2.isnull().sum())
print(hr_data_2.head())
```

以下输出结果显示，各属性的缺失值已被替换为均值：

```
satisfaction_level       0
last_evaluation          0
number_project           0
average_montly_hours     0
time_spend_company       0
Work_accident            0
left                     0
promotion_last_5years    0
sales                    0
salary                   0
dtype: int64
   satisfaction_level  last_evaluation  number_project  average_montly_hours  \
0                0.38             0.53        4.144477            157.000000
1                0.80             0.86        5.000000            200.698853
2                0.11             0.88        7.000000            272.000000
3                0.72             0.87        5.000000            223.000000
4                0.37             0.52        4.144477            159.000000

   time_spend_company  Work_accident  left  promotion_last_5years   sales  \
0                   3              0     1                    1.0   sales
1                   6              0     1                    1.0   sales
2                   4              0     1                    1.0   sales
3                   5              0     1                    1.0   sales
4                   3              0     1                    1.0   sales

   salary
0     low
1  medium
2  medium
3     low
4     low
```

（4）**使用指示变量**（using an indicator variable）：也可以生成一个二值变量，标识记录中是否有缺失值。再扩展到多个属性中，为每种属性创建二值指示变量。同样，也可以填充缺失值，再构建二值指示变量指明该值是真实值还是填充值。如果是故意空开的缺失值，结果也不会有偏差。

跟其他填充方法一样，先复制一份原始数据，创建新列代表填充的属性和值。以下代码首先新建列，在包含缺失值的属性的原列名后附加了_was_missing。

接着使用全局常量-999替换缺失值。这里使用的是全局常量填充法，但其实任何填充法都是可以的：

```
# make copy to avoid changing original data (when Imputing)
hr_data_4 = hr_data.copy()
# make new columns indicating what is imputed
cols_with_missing = (col for col in hr_data_4.columns
if hr_data_4[col].isnull().any())
for col in cols_with_missing:
    hr_data_4[col + '_was_missing'] = hr_data_4[col].isnull()
hr_data_4.fillna(-999, inplace=True)
hr_data_4.head()
```

由以下结果可见，新建了列，指标属性中是否有缺失值：

	Work_accident	left	promotion_last_5years	sales	salary	number_project_was_missing	average_montly_hours_was_missing	promotion_last_5years_was_missing
	0	1	-999.0	sales	low	True	False	True
	0	1	-999.0	sales	medium	False	True	True
	0	1	-999.0	sales	medium	False	False	True
	0	1	-999.0	sales	low	False	False	True
	0	1	-999.0	sales	low	True	False	True

（5）**使用数据挖掘算法预测最大概率值**（use a data mining algorithm to predict the most probable value）：使用机器学习算法，如KNN、线性回归、随机森林或决策树，预测缺失属性的最大概率值。这种方法的不足是，如果要在相同的数据集中用同样的算法完成其他预测或分类任务，则很可能出现数据过拟合。

在Python中，有一个fancyimpute库，这是一种可填充缺失值的先进数据挖掘方案。这是我们经常用到的一个库，所以应该在此演示一下。也有其他Python库能完成类似的任务。首先，使用以下命令安装fancyimpute库。必须使用命令提示符执行：

```
pip install fancyimpute
```

安装之后，返回到Jupyter notebook中，导入fancyimpute库中的KNN方法。

KNN 填充法只适用于数值。因此，首先从 hr_data 集中挑选出只有数值的列。然后，创建 KNN 模型，使 *k* 等于 3，替换数值属性中的缺失值：

```
from fancyimpute import KNN

hr_data_5 = hr_data.copy()
hr_numeric = hr_data_5.select_dtypes(include=[np.float])
hr_numeric = pd.DataFrame(KNN(3).complete(hr_numeric))
hr_numeric.columns = hr_numeric.columns
hr_numeric.index = hr_numeric.index
hr_numeric.head()
```

fancyimpute 库后端用的是 TensorFlow，执行需要一些时间。执行完成后，可滑动到下方查看填充结果，结果截图如下：

```
Imputing row 14201/14999 with 1 missing, elapsed time: 49.231
Imputing row 14301/14999 with 1 missing, elapsed time: 49.286
Imputing row 14401/14999 with 2 missing, elapsed time: 49.357
Imputing row 14501/14999 with 2 missing, elapsed time: 49.448
Imputing row 14601/14999 with 2 missing, elapsed time: 49.505
Imputing row 14701/14999 with 1 missing, elapsed time: 49.560
Imputing row 14801/14999 with 2 missing, elapsed time: 49.638
Imputing row 14901/14999 with 2 missing, elapsed time: 49.691
```

Out[14]:		satisfaction_level	last_evaluation	number_project	average_montly_hours	promotion_last_5years
	0	0.38	0.53	3.000000	157.000000	1.0
	1	0.80	0.86	5.000000	215.666667	1.0
	2	0.11	0.88	7.000000	272.000000	1.0
	3	0.72	0.87	5.000000	223.000000	1.0
	4	0.37	0.52	3.494849	159.000000	1.0

异常值

异常值（outliers）是与整体数据模式不符的极端值。异常值通常距离其他观测值非常远，使数据整体分布都扭曲了。建模过程输入异常值会导致错误的结果，因此合理地处理异常值是很有必要的。异常值一般可能分两种类型：单变量异常值和多变量异常值。

单变量异常值的检测和处理

顾名思义，单变量异常值基于数据集的单一属性。通过箱线图，或观测属性值的分布情况，就可识别出单变量异常值。但是，在构建 AutoML 流水线的过程中还没有机会看到数据分布。那么，AutoML 系统本身就要能检测并处理异常值。

因此，可部署以下任意一种方法，自动化进行单变量异常值的检测和处理。

- 四分位距和筛选（interquartile range and filtering）。
- 尾处理（winsorizing）。
- 裁切（trimming）。

创建一个伪异常值数据集，用于演示异常值的检测和处理方法：

```
%matplotlib inline
import numpy as np
import matplotlib.pyplot as plt
number_of_samples = 200
outlier_perc = 0.1
number_of_outliers = number_of_samples - int ( (1-outlier_perc) * number_of_samples )
# Normal Data
normal_data = np.random.randn(int ( (1-outlier_perc) * number_of_samples ),1)
# Inject Outlier data
outliers = np.random.uniform(low=-9,high=9,size=(number_of_outliers,1))
# Final data set
final_data = np.r_[normal_data,outliers]
```

使用以下代码对新建数据集绘图：

```
#Check data
plt.cla()
plt.figure(1)
plt.title("Dummy Data set")
plt.scatter(range(len(final_data)),final_data,c='b')
```

如图 3-4 所示，数据集尾部有一些异常值。

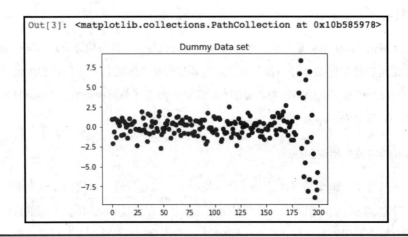

图 3-4　伪数据集

也可以使用如下代码生成箱线图来观测异常值。箱线图，也称**盒须图**（box and whisker），是一种根据 5 个数值表示数据分布的方法，5 个值为最小值、第一四分位、第二四分位、第三四分位和最大值。任何低于最小值和高于最大值的值都被视为异常值：

```
## Detect Outlier###
plt.boxplot(final_data)
```

从图 3-5 可以看出，有些值是在最大值和最小值之外的。那么，我们可以假设已经成功构造了一些异常值。

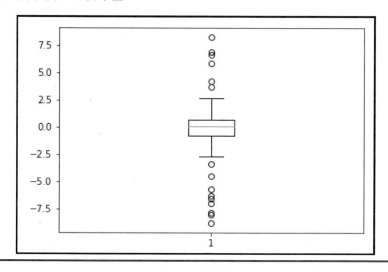

图 3-5　箱线图

排除异常值的一种方法是筛选出在最大值之上和最小值之下的值。要完成这项任务，需要计算**四分位间距**（inter-quartile range，IQR）。

四分位间距

四分位间距（IQR）是数据集可变性或分散性的度量值。把数据集分成四份，可计算出四分位间距。四分位将数据集前面提到的 5 个数分成四份，即最小值、第一四分位、第二四分位、第三四分位和最大值。第二四分位是排序数据集的中值；第一四分位是排序数据集第一半的中值，第三四分位是排序数据集第二半的中值。

四分位间距（IQR）是第三四分位（quartile75 或 Q3）和第一四分位（quartile25 或 Q1）之间的差。

使用如下 Python 代码计算 IQR：

```
## IQR Method Outlier Detection and Removal(filter) ##
quartile75, quartile25 = np.percentile(final_data, [75 ,25])
## Inter Quartile Range ##
IQR = quartile75 - quartile25
print("IQR",IQR)
```

从以下代码可见数据集的 IQR 为 1.49：

```
IQR 1.4941959696670106
```

筛选值

我们可以将大于最大值的值和小于最小值的值筛选出来。最小值可使用 quartile25-(IQR*1.5) 公式计算得出，最大值可使用 quartile75+(IQR*1.5) 公式计算得出。

最大值和最小值是依据 John Turkey 开发的 Turkey Fences 计算出来的。1.5 表示约有 1% 的异常值，与 3σ 同义，3σ 在许多统计测试中是作为边界用的。我们也可以选择 1.5 之外的其他值。但是，这个边界会增加或减少数据集中异常值的数量。

使用以下 Python 代码计算出数据集的 Max 值和 Min 值：

```
## Calculate Min and Max values ##
min_value = quartile25 - (IQR*1.5)
max_value = quartile75 + (IQR*1.5)
print("Max", max_value)
print("Min", min_value)
```

执行以上代码，输出如下。最大值和最小值分别为 2.94 和 -3.03：

```
Max 2.942189924241621
Min -3.034593954426421
```

接着，使用以下代码筛选出 min_value 以下的值和 max_value 以上的值：

```
filtered_values = final_data.copy()
filtered_values[ filtered_values< min_value] = np.nan
filtered_values[ filtered_values > max_value] = np.nan
#Check filtered data
plt.cla()
```

```
plt.figure(1)
plt.title("IQR Filtered Dummy Data set")
plt.scatter(range(len(filtered_values)),filtered_values,c='b')
```

代码执行成功后，就会看到图 3-6 中的异常值消失，数据集比之前好多了。

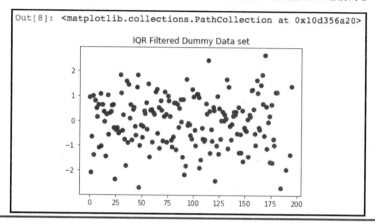

图 3-6　IQR 筛选后的伪数据集

尾处理

尾处理（winsorizing）是使用较小绝对值替换极端值的方法。把数值型列中的非零数值进行排序，计算尾部值，然后用所定义的参数替换尾部的值。

我们可以使用 SciPy 包中的 winsorize 方法处理异常值。SciPy 是一个 Python 库，集合了科学和技术计算空间相关的开源 Python 贡献。其中收集了大量的统计计算模块、线性代数、优化、信号与图像处理和许多其他模块。

导入 winsorize 方法后，将 data 和 limit 参数传入函数中。尾部值的计算和替换均由此方法实现，生成数据中不含缺失值：

```
##### Winsorization ####
from scipy.stats.mstats import winsorize
import statsmodels.api as sm
limit = 0.15
winsorized_data = winsorize(final_data,limits=limit)
#Check winsorized data
plt.cla()
plt.figure(1)
plt.title("Winsorized Dummy Data set")
plt.scatter(range(len(winsorized_data)),winsorized_data,c='b')
```

从图 3-7 中可看到，已经对极端值进行了尾处理，数据看起来没有异常值了。

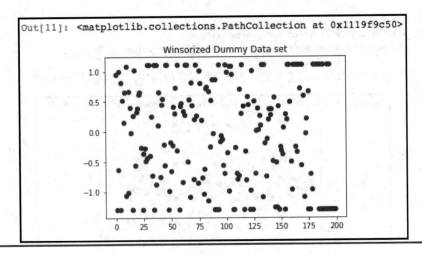

图 3-7 尾处理后的伪数据集

裁切

裁切（trimming）是与尾处理相同的方法，唯一的区别是尾部值是裁切出来的。

stats 库中的 trimboth 方法是从数据集中切掉两端的数据。把 final_data 和 0.1 限制参数传入函数中，裁切掉两端各 10% 的数据：

```
### Trimming Outliers ###
from scipy import stats
trimmed_data = stats.trimboth(final_data, 0.1)
#Check trimmed data
plt.cla()
plt.figure(1)
plt.title("Trimmed Dummy Data set")
plt.scatter(range(len(trimmed_data)),trimmed_data,c='b')
```

从图 3-8 中可看到极端值已被切掉，数据集中不再有异常值了。

多变量异常值的检测和处理

多变量异常值是至少两个变量的极端值的混合。单变量异常值检测方法很适合处理单维数据，但超过一维后，检测异常值就不太容易了。多变量异常值检测方法也是一种异常检测方法。检测多变量异常值可用到的技术有单类 SVM、**本地异常元素**（local outlier factor，LOF）和 IsolationForest。

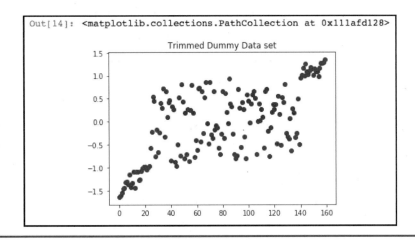

图 3-8　裁切后的伪数据集

我们使用以下的 IsolationForest 代码在人力流失数据集上演示多变量异常值的检测。从 sklearn.ensemble 的包中导入 IsolationForest。下一步，加载数据，将类别型变量转换为独热编码变量，使用估算器数量调用 IsolationForest 方法：

```
##Isolation forest
import numpy as np
import pandas as pd
from sklearn.ensemble import IsolationForest
hr_data = pd.read_csv('data/hr.csv', header=0)
print('Total number of records ',hr_data.shape)
hr_data = hr_data.dropna()
data_trnsf = pd.get_dummies(hr_data, columns =['salary', 'sales'])
data_trnsf.columns
clf = IsolationForest(n_estimators=100)
```

然后，将 IsolationForest 实例（clf）拟合到数据中，使用 predict 方法预测异常值。其中异常值用 -1 表示，非异常值（也称**新数据**（novel data））用 1 表示：

```
clf.fit(data_trnsf)
y_pred_train = clf.predict(data_trnsf)
data_trnsf['outlier'] = y_pred_train
print('Number of outliers ',data_trnsf.loc[data_trnsf['outlier'] == -1].shape)
print('Number of non outliers ',data_trnsf.loc[data_trnsf['outlier'] == 1].shape)
```

以下的输出结果显示，此模型可从数据集的 14999 个记录中识别出约 1500 个异常值：

```
Total number of records  (14999, 10)
Number of outliers       (1500, 22)
Number of non outliers   (13499, 22)
```

分箱

分箱（binning）是一个把连续数值分成较小量的桶或箱的过程。这是连续数据值离散化的一种重要技术。许多算法，如朴素贝叶斯（Naive Bayes）和 Apriori，都很适合离散数据集。因此，有必要将连续数值转换为离散值。

分箱方法有很多种。

- **等宽分箱**（equiwidth binning）：把数据分成k等份的等宽箱：
$$w = (maxval - minval)/k$$

其中：w是箱的宽度，$maxval$是数据的最大值，$minval$是数据的最小值，k是所求的箱数。

各个区间边界如下：
$$minval + w, minval + 2w, \ldots minval + (k-1)w$$

- **等频分箱**（equifrequency binning）：将数据分成k组，每组有相同数据量的值。

两种方法中，k的值是根据具体需求而定的，也是在试错过程中总结出来的。

除了这两种方法，我们也可明确指定切割点，创建分箱。当已知数据并想以某种形式分箱时，这种方法就非常有用。以下代码是依据预定义切割点进行分箱的一个函数：

```python
#Bin Values:
def bins(column, cutpoints, labels=None):
 #Define min and max values:
 min_val = column.min()
 max_val = column.max()
 print('Minimum value ',min_val)
 print(' Maximum Value ',max_val)
 break_points = [min_val] + cut_points + [max_val]
 if not labels:
    labels = range(len(cut_points)+1)
 #Create bins using the cut function in pandas
 column_bin =
pd.cut(column,bins=break_points,labels=labels,include_lowest=True)
```

```
    return column_bin
```

以下代码将员工满意度分成三个类别：low、medium 和 high。任何低于 0.3 的值都被视为 low 满意度，高于 0.6 的值被认为是 high 满意度，两个值之间为 medium：

```
import pandas as pd
hr_data = pd.read_csv('data/hr.csv', header=0)
hr_data.head()
hr_data = hr_data.dropna()
print(hr_data.shape)
print(list(hr_data.columns))
#Binning satisfaction level:
cut_points = [0.3,0.6]
labels = ["low","medium","high"]
hr_data["satisfaction_level"] = bins(hr_data["satisfaction_level"],
cut_points, labels)
print('\n####The number of values in each bin
are:###\n\n',pd.value_counts(hr_data["satisfaction_level"], sort=False))
```

执行以上代码，可看到如下结果，satisfaction_level 属性下创建了三个分箱，low 箱中有 1941 个值，medium 箱中有 4788 个值，high 箱中有 8270 个值：

```
(14999, 10)
['satisfaction_level', 'last_evaluation', 'number_project', 'average_montly_hours', 'time_spend_company', 'Work_accid
ent', 'left', 'promotion_last_5years', 'sales', 'salary']
0.09
1.0

####The number of values in each bin are:###

 low       1941
 medium    4788
 high      8270
Name: satisfaction_level, dtype: int64
```

对数和幂转换

对数和幂转换（log and power transformation）可减小高扭曲分布的扭曲度，有利于非树模型的构建。这一预处理技术有助于满足线性回归模型的假设和推论统计的假设。一些转换示例包括对数转换、方差转换和对数-对数转换。

使用伪数据集演示平方根转换，代码如下：

```
import numpy as np
values = np.array([-4, 6, 68, 46, 89, -25])
# Square root transformation #
sqrt_trnsf_values = np.sqrt(np.abs(values)) * np.sign(values)
print(sqrt_trnsf_values)
```

以上平方根转换输出如下:

```
[-2.         2.44948974 8.24621125 6.78232998 9.43398113 -5.        ]
```

接下来使用另一伪数据集尝试对数转换:

```
values = np.array([10, 60, 80, 200])
#log transformation #
log_trnsf_values = np.log(1+values)
print(log_trnsf_values)
```

伪数据集对数转换生成结果如下:

```
[2.39789527 4.11087386 4.39444915 5.30330491]
```

对数值型数据的不同预处理方法有了基本了解后,下面就来看看有哪些预处理类别型数据的方法。

类别型数据转换
Categorical Data Transformation

类别型数据本质上是非参数性的,也就意味着这些数据不会遵循任何分布。但是,在参数模型中使用这些变量的话,就需要通过各种编码方法对类别型数据进行转换,替换缺失值,且可用分箱技术减少类别的数量。

编码

许多机器学习实践中,数据集会包含类别类变量。这种数据尤其适合企业环境,因为企业的大多数属性是分类别的。这些变量有不同的离散值。比如,一个组织的规模可以是 Small、Medium 或 Large,地理位置可以是 Americas、Asia Pacific 和 Europe 等。许多机器学习算法,尤其是树模型,可直接处理类别型数据。

但是,许多算法不直接处理类别型数据。因此,如果要进一步处理,就要将这些属性编码成数值。类别型数据编码方法很多,一些经常用到的方法如下。

- 标签编码 (label encoding): 顾名思义,标签编码将类别标签转换成数值标签。标签编码更适合有序类别型数据。标签通常是 0 到 n−1 之间的值,n 是类的数量。

- **独热编码**（one-hot encoding）：也称伪编码（dummy encoding）。这种方法为类别属性/预测器的每一类都生成伪列。对每个伪预测器，有值用 1 表示，没有值用 0 表示。
- **基于频率编码**（frequency-based encoding）：本方法中，首先计算每个类的频率。然后计算每个类与总类的相对频率。把相对频率分配为属性类别的编码值。
- **目标均值编码**（target mean encoding）：本方法中，类别预测器的每个类都可编码为一个目标均值函数。此方法只能用于有目标特征的监督学习。
- **二值编码**（binary encoding）：首先把类转换成数值。然后把数值变成相似的二进制字符串。接着会分成不同的列。每个二进制位变成一个独立的列。
- **哈希编码**（hash encoding）：通常也称特征哈希。大多数人都知道有一个哈希函数可把数据映射成数值。这种方法中，同一个桶可能会被赋予不同的类，但如果一个输入特征中有几百个类别或类时，就很适合用哈希编码。

这些技术大多数都可以用 Python 实现，包含在 category_encoders 包里。使用以下命令安装 category_encoders 库：

```
pip install category_encoders
```

接下来导入 category_encoders 库为 ce（缩写代码，方便使用）。加载人力流失数据集，并为 salary 属性进行独热编码：

```
import pandas as pd
import category_encoders as ce
hr_data = pd.read_csv('data/hr.csv', header=0)
hr_data.head()
hr_data = hr_data.dropna()
print(hr_data.shape)
print(list(hr_data.columns))
onehot_encoder = ce.OneHotEncoder(cols=['salary'])
onehot_df = onehot_encoder.fit_transform(hr_data)
onehot_df.head()
```

可以看到，很轻松就能使用 category_encoders 库把类别属性转换成对应的独热编码属性，如图 3-9 所示。

```
Out[29]:
         salary_0  salary_1  salary_2  salary_-1
      0     1        0         0          0
      1     0        1         0          0
      2     0        1         0          0
      3     1        0         0          0
      4     1        0         0          0
```

图 3-9 独热编码转换

类似地，使用 OrdinalEncoder 可对 salary 数据进行标签编码：

```
ordinal_encoder = ce.OrdinalEncoder(cols=['salary'])
ordinal_df = ordinal_encoder.fit_transform(hr_data)
ordinal_df.head(10)
ordinal_df['salary'].value_counts()
```

以上代码将 low、medium 和 high 薪酬类别映射到三个数值 0、1、2 上：

```
Out[30]:  0    7316
          1    6446
          2    1237
          Name: salary, dtype: int64
```

类似地，使用以下代码尝试 CategoryEncoders 中的其他类别编码方法，观察结果：

```
binary_encoder = ce.BinaryEncoder(cols=['salary'])
df_binary = binary_encoder.fit_transform(hr_data)
df_binary.head()

poly_encoder = ce.PolynomialEncoder(cols=['salary'])
df_poly = poly_encoder.fit_transform(hr_data)
df_poly.head()
helmert_encoder = ce.HelmertEncoder(cols=['salary'])
helmert_df = helmert_encoder.fit_transform(hr_data)
helmert_df.head()
```

下一个话题是类别属性缺失值的处理方法。

类别型数据转换的缺失值

评估类别变量的缺失值的技术也是一样的。但是，有些填充技术不同，也有些与之前讨论过的数值缺失值的处理方法类似。下面演示只处理类别缺失值的 Python 代码。

- **去除或删除数据**（remove or delete the data）：决定是否删除类别变量中缺失的数据点的过程与以前数值缺失值处理的过程是一样的。
- **使用众数替换缺失值**（replace missing values with the mode）：因为类别型数据是非参数的，与数值不同，所以类别型数据没有均值或中值。因此，最简单的方法是使用众数替换类别缺失值。众数是类别变量中最高频出现的类。比如，假设有一个有三个类红、绿和蓝的预测器。红在数据集中最常发生，频率为 30，绿发生频率为 20，蓝发生频率为 10。那么，缺失值就可以用红替代，因为红是预测器中最高频发生的。

使用人力流失数据集来演示类别属性的缺失值处理的过程。首先将数据集加载进来，观察数据集中空值的数量：

```
import numpy as np
import pandas as pd
hr_data = pd.read_csv('data/hr.csv', header=0)
print('Nulls in the data set' ,hr_data.isnull().sum())
```

从以下的输出中可以看到，类别属性 sales 和 salary 中没有缺失值。所以，就要为这些特征制造一些缺失值：

```
Nulls in the data set satisfaction_level    0
last_evaluation          0
number_project           0
average_montly_hours     0
time_spend_company       0
Work_accident            0
left                     0
promotion_last_5years    0
sales                    0
salary                   0
dtype: int64
```

使用以下代码将 sales 属性中的 sales 值替换为空值，将 salary 属性的 low 值替换为空值：

```
#As there are no null introduce some nulls by replacing sales in sales
column with NaN
hr_data[['sales']] = hr_data[[ 'sales']].replace('sales', np.NaN)
#As there are no null introduce some nulls by replacing low in salary
column with NaN
hr_data[['salary']] = hr_data[[ 'salary']].replace('low', np.NaN)
print('New nulls in the data set' ,hr_data.isnull().sum())
```

执行代码，就可在 salary 和 sales 属性中发现空值，输出如下：

```
New nulls in the data set satisfaction_level    0
last_evaluation          0
number_project           0
average_montly_hours     0
time_spend_company       0
Work_accident            0
left                     0
promotion_last_5years    0
sales                 4140
salary                7316
dtype: int64
```

现在可将这些空值替换为每列的众数。与数值缺失值填充步骤一样，先创建一份 hr_data 复本，保留好原始数据集。接着使用 fillna 方法用众数填充各行，代码如下所示：

```
#Replace mode for missing values
hr_data_1 = hr_data.copy()
# fill missing values with mode column values
for column in hr_data_1.columns:
 hr_data_1[column].fillna(hr_data_1[column].mode()[0], inplace=True)
# count the number of NaN values in each column
print(hr_data_1.isnull().sum())
print(hr_data_1.head())
```

输出如下所示，sales 列中缺失值已经替换成 technical，salary 列已经替换成 medium：

```
satisfaction_level       0
last_evaluation          0
number_project           0
average_montly_hours     0
time_spend_company       0
Work_accident            0
left                     0
promotion_last_5years    0
sales                    0
salary                   0
dtype: int64
   satisfaction_level  last_evaluation  number_project  average_montly_hours  \
0                0.38             0.53               2                   157
1                0.80             0.86               5                   262
2                0.11             0.88               7                   272
3                0.72             0.87               5                   223
4                0.37             0.52               2                   159

   time_spend_company  Work_accident  left  promotion_last_5years       sales  \
0                   3              0     1                      0   technical
1                   6              0     1                      0   technical
2                   4              0     1                      0   technical
3                   5              0     1                      0   technical
4                   3              0     1                      0   technical

    salary
0   medium
1   medium
2   medium
3   medium
4   medium
```

- 使用全局常量填充缺失值（use a global constant to fill in the missing value）：
与数值缺失值的处理方法相同，可使用全局常量，如 AAAA 或 NA，区分缺失值与数据集其他值：

```
#Mark global constant for missing values
hr_data_2 = hr_data.copy()
# fill missing values with global constant values
hr_data_2.fillna('AAA', inplace=True)
# count the number of NaN values in each column
print(hr_data_2.isnull().sum())
print(hr_data_2.head())
```

- 使用指示变量（using an indicator variable）：与数值型变量相似，可以为缺失类别型数据填充值添加一个指标变量进行标识。
- 使用数据挖掘算法预测最大概率值（use a data mining algorithm to predict the most probable value）：如数值型属性一样，也可以使用数据挖掘算法，如决策树、随机森林或 KNN 方法来预测缺失值的最大概率值。此任务也可用同一个 fancyimpute 库完成。

我们讨论了结构化数据的预处理方法。在这个数字化时代中，也会有许多来自各处的非结构化数据。后面章节中我们会了解一些预处理文本数据的方法，便于建模使用。

文本预处理
Text Preprocessing

分析时，很有必要去除那些平添干扰的冗余文本，以减小文本数据的特征空间大小。文本数据预处理通常会有一系列步骤。然而，并非每个任务都要执行全部步骤，按需执行即可。比如，若一个文本数据中的每个单词都已经小写，就不用再统一文本大小写。

一个文本预处理任务有三个主要成分。

- 标记化（tokenization）。
- 归一化（normalization）。

- 替换（substitution）。

我们使用 nltk 库演示不同的文本预处理方法。执行以下命令，安装 nltk 库：

```
pip install nltk
```

安装完成后，在 Python 环境中运行如下代码：

```
##Run this cell only once##
import nltk
nltk.download()
```

就会得到一个 NLTK 下载器的弹窗。在标识栏中选择全部，等待安装完成。

本节我们会学习文本预处理的步骤，实现文本归一化。

（1）标记化（tokenization）：这种方法将文本分成较小的块，如句子或单词。同时，一些文本挖掘任务，如 Word2Vec 模型准备，会更偏向将文本分成段落或句子。因此，我们可以使用 NLTK 的 sent_tokenize 将文本转换成句子。首先，使用以下代码从 data 文件夹中读取文本文件：

```
import pandas as pd
import category_encoders as ce
text_file = open('data/example_text.txt', 'rt')
text = text_file.read()
text_file.close()
```

（2）将文本标记为句子，从 nltk 库中导入 sent_tokenize 方法，将文本作为参数传入：

```
## Sentence tokenization ##
from nltk import sent_tokenize
sentence = sent_tokenize(text)
print(sentence[0])
```

以上代码输出如下：

```
Data Science sits at the core of any analytical exercise conducted on a Big Data or Internet of Things (IoT) environment.
```

（3）相似地，一些建模方法，如词袋模型，需要把文本转换为独立的单词形式。本例中可以使用 NLTK 的 word_tokenize 方法将文本转换成单词，代码如下所示：

```
## Word tokenization ##
from nltk import word_tokenize
```

```
words = word_tokenize(text)
print(words[:50])
```

（4）以下输出结果显示从文本中提取出 50 个标记化单词。可以发现，有一些非字母字符，如标点符号，也被标记化了。这对分析活动没有任何价值，因此需要将这些变量移除：

```
['Data', 'Science', 'sits', 'at', 'the', 'core', 'of', 'any', 'analytical', 'exercise', 'conducted', 'on', 'a', 'Big', 'Data', 'or', 'Internet', 'of', 'Things', '(', 'IoT', ')', 'environment', '.', 'Data', 'science', 'involves', 'a', 'wide', 'array', 'of', 'technologies', ',', 'business', ',', 'and', 'machine', 'learning', 'algorithms', '.', 'The', 'purpose', 'of', 'data', 'science', 'is', 'just', 'not', 'doing', 'machine']
```

（5）准备词袋模型的时候，非字母字符，如标点符号，不具有任何价值，所以各种标点和符号，如、""、+和-等，均可删除。

有许多移除非字母字符的方法。下面展示一个 Python 的方法：

```
# Remove punctuations and keep only alphabets
words_cleaned = [word for word in words if word.isalpha()]
print(words_cleaned[:50])
```

（6）可从如下标记中发现，无关的字符，如(之类的，已经从标记列表中移除。但是，一些常用的单词，如 at 和 of，对分析也没有价值，可采用停止词移除法删除：

```
['Data', 'Science', 'sits', 'at', 'the', 'core', 'of', 'any', 'analytical', 'exercise', 'conducted', 'on', 'a', 'Big', 'Data', 'or', 'Internet', 'of', 'Things', 'IoT', 'environment', 'Data', 'science', 'involves', 'a', 'wide', 'array', 'of', 'technologies', 'business', 'and', 'machine', 'learning', 'algorithms', 'The', 'purpose', 'of', 'data', 'science', 'is', 'just', 'not', 'doing', 'machine', 'learning', 'or', 'statistical', 'analysis', 'but', 'also']
```

（7）停止词是编写文本文件常用到的短功能词，可能是填充语或介词。NLTK 提供了一个标准的英语停止词集，可用来过滤掉文本中的停止词。同样，有时一些特定领域的停止词也可用于除去非正式单词。我们随时可以从文本中创建一个单词列表，列出我们认为与分析无关的单词。

要移除停止词，首先要从 nltk.corpus 库中导入 stopwords 方法。然后从 stopwords.words 方法中调用英语停止词词典，去除在标记列表中发现的常用词：

```
# remove the stop words
from nltk.corpus import stopwords
stop_words = set(stopwords.words('english'))
words_1 = [word for word in words_cleaned if not word in stop_words]
print(words_1[:50])
```

(8）如下输出可见，此前出现的单词，如 at 和 the 等已经从标记列表中移除。但是，有一些相似的单词，如 Data 和 data 在标记列表中却分为两个不同的单词出现了。所以要将这些词转换成相同的大小写格式：

```
['Data', 'Science', 'sits', 'core', 'analytical', 'exercise', 'conducted', 'Big', 'Data', 'Internet', 'Things', 'IoT', 'environment', 'Data', 'science', 'involves', 'wide', 'array', 'technologies', 'business', 'machine', 'learning', 'algorithms', 'The', 'purpose', 'data', 'science', 'machine', 'learning', 'statistical', 'analysis', 'also', 'derive', 'insights', 'data', 'user', 'statistics', 'knowledge', 'understand', 'In', 'fast', 'paced', 'environment', 'Big', 'Data', 'IoT', 'type', 'data', 'might', 'vary']
```

（9）大小写统一转换（case folding）是将全部单词转换到统一大小写格式，不区分大小写。通常将全部大写字母转换成小写字母。

可使用 lower 函数将全部大写字母转换成小写字母，如下代码所示：

```
# Case folding
words_lower = [words_1.lower() for words_1 in words_1]
print(words_lower[:50])
```

从输出的结果中可以看到 Data 这样的词不再出现在列表中了，全部转换成小写字母了：

```
['data', 'science', 'sits', 'core', 'analytical', 'exercise', 'conducted', 'big', 'data', 'internet', 'things', 'iot', 'environment', 'data', 'science', 'involves', 'wide', 'array', 'technologies', 'business', 'machine', 'learning', 'algorithms', 'the', 'purpose', 'data', 'science', 'machine', 'learning', 'statistical', 'analysis', 'also', 'derive', 'insights', 'data', 'user', 'statistics', 'knowledge', 'understand', 'in', 'fast', 'paced', 'environment', 'big', 'data', 'iot', 'type', 'data', 'might', 'vary']
```

词根提取（stemming）是让单词回归到其基础或根形式。比如，worked 和 working 都是由 to work 衍生出来的。将全部相似的单词转换成其基础形式是很有用的，可以促进更好的情感分析、文件分类，以及许多研究。

从 nltk.stem.porter 库中导入 PorterStemmer，将 PorterStemmer 类实例化。接着将 words_lower 标记列表传入 porter.stem 类中，将每个词都提取为根形式：

```
#Stemming
from nltk.stem.porter import PorterStemmer
porter = PorterStemmer()
stemmed_words = [porter.stem(word) for word in words_lower]
print(stemmed_words[:50])
```

以上代码生成如下标记的根提取列表：

```
['data', 'scienc', 'sit', 'core', 'analyt', 'exercis', 'conduct', 'big', 'data', 'internet', 'thing', 'iot', 'environ', 'data', 'scienc', 'involv', 'wide', 'array', 'technolog', 'busi', 'machin', 'learn', 'algorithm', 'the', 'purpos', 'data', 'scienc', 'machin', 'learn', 'statist', 'analysi', 'also', 'deriv', 'insight', 'data', 'user', 'statist', 'knowledg', 'understand', 'in', 'fast', 'pace', 'environ', 'big', 'data', 'iot', 'type', 'data', 'might', 'vari']
```

并不是所有特征或属性都是机器学习模型中的重要因素。下面会了解在准备机器学习流水线的时候减少特征数量的一些方法。

3.3 特征选择
Feature Selection

机器学习模型中采用一些关键特征学习数据中的模式。其他非关键特征只会增加对模型的干扰，可能导致模型的准确率下降以及过度拟合。因此，特征选择非常必要。而且精选出的核心特征会缩短模型训练的时间。

以下是创建模型前选择特征的一些方法。

- 识别关联变量，移除高度关联值。
- 低方差特征排除。
- 衡量特征集的信息增益，选择前N个特征。

同样，创建基线模型后，如下方法可辅助选出合适的特征。

- 采用线性回归并基于P值选择变量。
- 采用线性回归按步骤选择重要变量。
- 采用随机森林选择前N个重要变量。

下面介绍 scikit-learn 中一些精简数据集特征的方法。

低方差特征排除
Excluding Features with Low Variance

方差或可变性不大的特征对机器学习模型学习数据模式没有实用价值。比如，某数据集的一个特征是每条记录都有一个值是 5，是一个常量，那就不算重要特征，需要排除。

scikit-learn 的 feature_selection 模块中的 VarianceThreshold 方法用于去除方差

低于标准或阈值的特征。sklearn.feature_selection 模块执行特征选择算法。现在有单变量过滤选择法和递归特征去除法。以下示例演示 sklearn.feature_selection 如何使用：

```
%matplotlib inline
import pandas as pd
import numpy as np
from sklearn.feature_selection import SelectKBest
from sklearn.feature_selection import chi2
hr_data = pd.read_csv('data/hr.csv', header=0)
hr_data.head()
hr_data = hr_data.dropna()
data_trnsf = pd.get_dummies(hr_data, columns =['salary', 'sales'])
data_trnsf.columns
```

以上代码输出如下：

```
Out[10]: Index(['satisfaction_level', 'last_evaluation', 'number_project',
               'average_montly_hours', 'time_spend_company', 'Work_accident', 'left',
               'promotion_last_5years', 'salary_high', 'salary_low', 'salary_medium',
               'sales_IT', 'sales_RandD', 'sales_accounting', 'sales_hr',
               'sales_management', 'sales_marketing', 'sales_product_mng',
               'sales_sales', 'sales_support', 'sales_technical'],
              dtype='object')
```

接下来将 left 赋为目标变量，其他属性作为独立属性，如下代码所示：

```
X = data_trnsf.drop('left', axis=1)
X.columns
Y = data_trnsf.left# feature extraction
```

数据备好，采用 VarianceThreshold 方法选择特征。首先，从 scikit-learn 的 feature_seleciton 模块中导入 VarianceThreshold 方法。然后，在 VarianceThreshold 方法中设置 threshold 为 0.2，即属性数据方差小于 20%时，去除该属性，不选为特征。执行如下代码，观察精简后的特征集：

```
#Variance Threshold
from sklearn.feature_selection import VarianceThreshold
# Set threshold to 0.2
select_features = VarianceThreshold(threshold = 0.2)
select_features.fit_transform(X)
# Subset features
X_subset = select_features.transform(X)
print('Number of features:', X.shape[1])
print('Reduced number of features:',X_subset.shape[1])
```

输出如下，20 个属性中有 5 个方差大于 20%，具有可变性，通过方差阈值选择：

```
Number of features: 20
Reduced number of features: 5
```

下面学习单变量特征选择法,即根据统计学测试选出重要特征。

单变量特征选择
Univariate Feature Selection

本方法中分别对每个特征进行统计测试。根据测试结果选择最好的特征。

以下示例演示了使用 chi-squared 统计测试从人力流失数据集中选择最佳特征的过程:

```
#Chi2 Selector

from sklearn.feature_selection import SelectKBest
from sklearn.feature_selection import chi2

chi2_model = SelectKBest(score_func=chi2, k=4)
X_best_feat = chi2_model.fit_transform(X, Y)
# selected features
print('Number of features:', X.shape[1])
print('Reduced number of features:',X_best_feat.shape[1])
```

输出如下,选出 4 个特征。调整 k 值,可改变最佳特征的数量:

```
Number of features: 20
Reduced number of features: 4
```

下面演示递归特征消除法。

递归特征消除
Recursive Feature Elimination

递归特征消除的思路是反复建模,移除特征,用剩余特征建模,计算模型准确率,重复此步骤,直到遍历全部特征。这是一种找出最佳特征子集的贪婪优化法,然后根据被移除时间对特征进行排序。

以下示例代码中,使用人力流失数据集展示**递归特征消除**(recursive feature elimination,RFE)的用法。RFE 方法的稳定性在很大程度上依赖于所使用的算法。本例中采用了 LogisticRegression 方法:

```
#Recursive Feature Elimination
```

```python
from sklearn.feature_selection import RFE
from sklearn.linear_model import LogisticRegression

# create a base classifier used to evaluate a subset of attributes
logistic_model = LogisticRegression()

# create the RFE model and select 4 attributes
rfe = RFE(logistic_model, 4)
rfe = rfe.fit(X, Y)

# Ranking of the attributes
print(sorted(zip(map(lambda x: round(x, 4), rfe.ranking_),X)))
```

输出显示了排序后的特征：

```
[ 1  7  9 17  6  1  1  1  3  8 15  2 12  5  4 13 16 14 11 10]
[(1, 'Work_accident'), (1, 'promotion_last_5years'), (1, 'salary_high'), (1, 'satisfaction_level'), (2, 'sales_Rand
D'), (3, 'salary_low'), (4, 'sales_management'), (5, 'sales_hr'), (6, 'time_spend_company'), (7, 'last_evaluation'),
(8, 'salary_medium'), (9, 'number_project'), (10, 'sales_technical'), (11, 'sales_support'), (12, 'sales_accountin
g'), (13, 'sales_marketing'), (14, 'sales_sales'), (15, 'sales_IT'), (16, 'sales_product_mng'), (17, 'average_montly_
hours')]
```

机器学习流水线中经常使用随机森林法选择特征。因此，很有必要了解随机森林法。

随机森林特征选择
Feature Selection Using Random Forest

随机森林中用到的是树型特征选择策略，依据特征对节点纯度的贡献来对特征进行排序。首先，要构建随机森林模型。第2章中已经了解了随机森林模型的构建过程，代码如下：

```python
# Feature Importance
from sklearn.ensemble import RandomForestClassifier
# fit a RandomForest model to the data
model = RandomForestClassifier()
model.fit(X, Y)
# display the relative importance of each attribute
print(model.feature_importances_)
print(sorted(zip(map(lambda x: round(x, 4),
model.feature_importances_),X)))
```

模型构建成功后，使用feature_importance_attribute查看排序的特征，结果如下所示：

```
[(0.001, 'sales_product_mng'), (0.0012, 'sales_marketing'), (0.0014, 'promotion_last_5years'), (0.0015, 'sales_Rand
D'), (0.0015, 'sales_accounting'), (0.0017, 'sales_management'), (0.0019, 'sales_hr'), (0.002, 'sales_IT'), (0.0025,
'sales_support'), (0.0037, 'sales_technical'), (0.0038, 'salary_medium'), (0.0039, 'sales_sales'), (0.0071, 'salary_l
ow'), (0.0073, 'salary_high'), (0.0121, 'Work_accident'), (0.1179, 'last_evaluation'), (0.1187, 'average_montly_hour
s'), (0.1543, 'number_project'), (0.2152, 'time_spend_company'), (0.3413, 'satisfaction_level')]
```

本节中讨论了几种从全特征集中选出子集的方法。接下来看看特征选择的降维法。

降维特征选择
Feature Selection Using Dimensionality Reduction

降维法是合并原始特征，构建新特征，从而减少特征维度。这种方法有效保留了数据的可变性，但不足是新属性很难解释，因为新属性是由各种元素组合而成的。

主成分分析

主成分分析（principal component analysis，PCA）将数据从高维空间转换到低维空间。可以想象，若对一个有100个维度的数据集进行可视化，基本是不可能高效地展现出数据分布形状的。PCA 是合成主成分，从而有效降维的方法，可在低维空间诠释数据的可变性。

使用数学公式表示，已知一套变量 $X_1, X_2,....., X_p$，其中有 p 个原始变量。在 PCA 中，要找出一套新变量，$Z_1, Z_2,....., Z_p$，为原始变量（减去均值后）的权重均值：

$$Z_p = a_{i,1}(X_1 - \bar{X_1}) + a_{i,2}(X_2 - \bar{X_2}) +...+ a_{i,p}(X_p - \bar{X_p})$$

其中每一对Z的相关性=0。

所得到的Z按照方差排序，Z_1 方差最大，Z_p 方差最小。

> 每次从 PCA 提取出的第一个成分在整体预测变量总方差中占比最高。提取出的第二个成分是除去第一个成分之外的数据集方差中占比最高的，且与第一个成分无相关性。若计算第一个成分和第二个成分之间的相关性，结果会是 0。

使用人力流失数据演示 PCA 的用法。首先，在环境中导入 numpy 和 pandas 库，并加载人力数据集：

```python
import numpy as np
import pandas as pd
hr_data = pd.read_csv('data/hr.csv', header=0)
print (hr_data.head())
```

输出如下，显示数据集中每个属性的前 5 行：

```
   satisfaction_level  last_evaluation  number_project  average_montly_hours  \
0                0.38             0.53               2                   157
1                0.80             0.86               5                   262
2                0.11             0.88               7                   272
3                0.72             0.87               5                   223
4                0.37             0.52               2                   159

   time_spend_company  Work_accident  left  promotion_last_5years  sales  \
0                   3              0     1                      0  sales
1                   6              0     1                      0  sales
2                   4              0     1                      0  sales
3                   5              0     1                      0  sales
4                   3              0     1                      0  sales

   salary
0     low
1  medium
2  medium
3     low
4     low
```

PCA 很适合数值型属性，属性标准化时效果较好。那么，从 sklearn.preprocessing 库中导入 StandardScaler。数据预处理时只采用数值型属性，使用 StandardScaler 法将人力流失数据集的数值型属性标准化：

```python
from sklearn.preprocessing import StandardScaler
hr_numeric = hr_data.select_dtypes(include=[np.float])
hr_numeric_scaled = StandardScaler().fit_transform(hr_numeric)
```

下一步从 sklearn.decomposition 导入 PCA 方法，给 n_components 传参为 2。n_components 是指定待构建主成分数量的参数。然后，确定这两个主成分解释的方差：

```python
from sklearn.decomposition import PCA
pca = PCA(n_components=2)
principalComponents = pca.fit_transform(hr_numeric_scaled)
principalDf = pd.DataFrame(data = principalComponents,columns = ['principal component 1', 'principal component 2'])
print(pca.explained_variance_ratio_)
```

输出如下，两个主成分可展示人力数据集的数值型属性的变化：

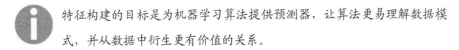

有时，当原始数据信息不足以构建好模型时，则需构建特征。接下来会讲解一些构建特征的方法。

3.4 特征生成
Feature Generation

基于已有特征构建新特征是一种艺术，有许多构建方法。

> 特征构建的目标是为机器学习算法提供预测器，让算法更易理解数据模式，并从数据中衍生更有价值的关系。

例如，在人力流失问题中，员工在职时长是重要属性。但是，有时数据集中没有在职时长这种属性，但有员工的入职时间。这种情况下，可以通过使用当前日期减去入职日期构建出在职时长数据。

以下是一些从已有数据中生成新特征的方法，但这并不全面，只是一些常用的方法。我们要仔细思考问题，探索数据，勇于创新，发掘构建特征的新方法。

- **数值型特征生成**（numerical feature generation）：从数值型数据生成新特征，比从其他类型数据生成或多或少都更容易些。数值型特征，即使不理解其中的含义，也可以执行各种运算，如将两个或多个数值相加、相减、相乘或相除。完成运算后，即可从生成的全部特征中挑出重要的特征，消除其余特征。虽然这是一个资源密集型任务，但对于无法直接衍生新特征的场景，借助此方法可发现新特征。

一对数值特征之间相加或相减，称为**按对特征创建**（pairwise feature creation）。

另一种方法称为 `PolynomialFeatures`（多项式特征）创建，自动进行特征的全部多项式组合。有利于映射特征之间表征某种独特状态的复杂关系。

scikit-learn 的 `PolynomialFeatures` 方法可用于生成多项式特征。先创建伪数据，代码如下所示：

```
#Import PolynomialFeatures
from sklearn.preprocessing import PolynomialFeatures
#Create matrix and vectors
var1 = [[0.45, 0.72], [0.12, 0.24]]
```

接着调用 `PolynomialFeatures`，定义参数级别，生成多项式特征。生成的特征级别比参数定义的级别低或同级：

```
# Generate Polynomial Features
ploy_features = PolynomialFeatures(degree=2)
var1_ = ploy_features.fit_transform(var1)
print(var1_)
```

代码执行完成后，生成新特征，如下所示：

```
[[1.     0.45    0.72    0.2025 0.324   0.5184]
 [1.     0.12    0.24    0.0144 0.0288  0.0576]]
```

- **类别特征创建**（categorical feature creation）：从类别型数据创建新特征的方法较有限，但可计算每个类别属性的频率，或组合不同的变量构建新特征。

- **时间特征创建**（temporal feature creation）：如果遇到日期/时间特征，则可以衍生出各种新特征，如下：

    ◆ 一周的哪一天；

    ◆ 一月的哪一天；

    ◆ 一季的哪一天；

    ◆ 一年的哪一天；

    ◆ 一天的哪一个小时；

    ◆ 一天的哪一秒；

    ◆ 一天的哪一周；

    ◆ 一年的哪一周；

◆ 一年的哪一月。

从日期/时间特征中创建出的新特征,有助于算法更好地学习数据的时间模式。

3.5 总结
Summary

本章中,我们学习了与机器学习流水线密切相关的数据转换和预处理方法。准备属性、清洗数据、清除数据误差,才可保障学习的正确性。排除数据中的干扰,生成优质的特征,有利于机器学习模型更高效地发现数据模式。

第4章的重点是AutoML算法。我们会讨论算法相关的特征转换,自动监督学习和无监督学习,以及其他扩展内容。

第 4 章

自动化算法选择
Automated Algorithm Selection

本章对机器学习算法进行全景概览。前面虽然已经介绍过，但再次简单回顾，从更宏观的角度了解哪些问题可通过机器学习处理。

数据集的样本/观测值之间若有关联标签，则在建模过程中对算法有指引作用。有指引或监督的场景，可采用监督或半监督学习算法。若没有标签，则可采用无监督学习算法。

也有一些例外，要求方法不同，如强化学习，但本章重点关注监督学习算法和无监督学习算法。

机器学习流水线的下一个前沿是自动化。初遇 AutoML 流水线，首先想到的核心环节是特征转换、模型选择和超参数优化。但是，具体问题中还需考虑其他因素，本章涉及如下几点。

- 计算复杂度。
- 训练时间和推理时间的区别。
- 线性与非线性。
- 算法所需的特征转换。

理解这些因素后，会更容易针对某个特定问题匹配合适的算法。本章结束时：

- 你会学到自动监督学习和无监督学习的基础知识。
- 你会学到使用机器学习流水线的主要考虑因素。
- 你可练习各种样例，创建监督机器学习流水线和无监督机器学习流水线。

4.1 技术要求
Technical Requirements

在 GitHub 的 Chapter 04 中查看 requirements.txt 文件，有运行本章示例代码要安装的库。

全部代码示例都在 GitHub 上的 Chapter 04 文件夹中。

4.2 计算复杂度
Computational Complexity

计算效率和复杂度是选择机器学习的重要考量因素，因为关系到模型训练和评估所需的资源，即对时间和内存的要求。

比如，计算密集的算法中，模型训练和超参数优化要求的时长更长。可通过把工作负载分布到更多 CPU 或 GPU 上，以空间换时间，节省更多时间。

本节从限制条件角度来衡量一些算法，但在深入算法细节之前，我们先了解一些算法复杂度的基本情况。

 算法的复杂度取决于输入的大小。机器学习算法的输入是指元素和特征的数量。通常会把在最差的条件下完成任务所需的操作数量理解为算法的复杂度。

大 O 表示法
Big O Notation

大家可能听过大 O 表示法。大 O 符号由不同的类表示复杂度，如线性——

O(n)、对数——O(log n)、平方——O(n2)、立方——O(n3),以及其他类。采用大 O 表示法的原因是算法的运行时间高度依赖于硬件,并且基于输入大小测量算法性能也需要有系统的方法。大 O 表示法遍历算法的每个步骤并能算出上述最差条件下的复杂度。

例如,若 n 是欲附加到列表中的元素的数量,则其复杂度是 O(n),因为附加操作的数量取决于 n。以下代码可绘制出复杂度随着输入大小变化而变化的图表:

```python
# Importing necessary libraries
import pandas as pd
import numpy as np
import matplotlib.pyplot as plt
import seaborn as sns

# Setting the style of the plot
plt.style.use('seaborn-whitegrid')

# Creating an array of input sizes
n = 10
x = np.arange(1, n)

# Creating a pandas data frame for popular complexity classes
df = pd.DataFrame({'x': x,
                   'O(1)': 0,
                   'O(n)': x,
                   'O(log_n)': np.log(x),
                   'O(n_log_n)': n * np.log(x),
                   'O(n2)': np.power(x, 2), # Quadratic
                   'O(n3)': np.power(x, 3)}) # Cubic

# Creating labels
labels = ['$O(1) - Constant$',
          '$O(\log{}n) - Logarithmic$',
          '$O(n) - Linear$',
          '$O(n^2) - Quadratic$',
          '$O(n^3) - Cubic$',
          '$O(n\log{}n) - N \log n$']

# Plotting every column in dataframe except 'x'
for i, col in enumerate(df.columns.drop('x')):
    print(labels[i], col)
    plt.plot(df[col], label=labels[i])

# Adding a legend
plt.legend()

# Limiting the y-axis
plt.ylim(0,50)
```

```
plt.show()
```

以上代码输出如图 4-1 所示。

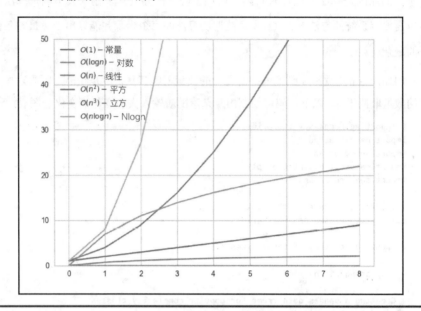

图 4-1　不同函数输入大小的复杂度变化

需注意，不同级别的复杂度之间会有交叉点。由此可以看出数据大小对复杂度的影响。简单样本的复杂度很容易理解，但机器学习算法的复杂度也容易理解吗？如果感兴趣，就请继续阅读下一节。

4.3　训练时间和推理时间的区别
Differences in Training and Scoring Time

训练时间和推理时间的消耗可以成就一个机器学习项目，也可毁灭一个机器学习项目。如果一种算法在当前可用的硬件上训练要花费的时间太长，进行数据更新和超参数优化就会非常痛苦，可能不得不放弃该算法。如果一种算法评估时间太长，那么很可能成为生产环境中的瓶颈，因为实际应用中要求快速推理，如在几毫秒或者几微秒内做出预测。所以，很有必要了解机器学习算法的内部工作机制，至少先了解常用算法的机制，从感官上初步判断算法是否合适。

比如，监督学习算法在训练过程中会了解样本集与其关联标签之间的关系，每个样本都包含一套特征集。训练任务完成后会输出一个机器学习模型，用来进行新预测。给模型输入不带标签的新样本时，基于训练期间学到的特征与标签的映射关系，为新样本预测标签。预测所需的时间通常很少，因为学到的模型权重会应用到新数据中。

但一些监督学习算法会跳过训练阶段，基于训练数据集中的全部可用样本进行推理。这种算法称为**基于实例**（instance-based）的算法或**懒惰学习者**（lazy learners）的算法。基于实例的算法，训练意味着把全部特征向量和关联标签存储到内存中，即整个训练数据集。这就说明，数据集越大，模型所需的计算量越大以及内存资源越多。

训练时间和推理时间的简化度量
Simple Measure of Training and Scoring Time

我们来看一个 K 近邻（KNN）算法的简单样本，适用于分类问题和回归问题。当一个算法为新特征向量评估时，会检查其 k 个最近邻居，然后输出结果。若是分类问题，则使用多数投票法做出预测；若是回归问题，则使用平均值做出预测。

下面练习一个分类问题，进一步理解这种方法。首先，创建一个样本数据集，检验 KNN 算法训练和推理消耗的时间。

为简化起见，使用如下函数测量特定命令所消耗的时间：

```
from contextlib import contextmanager
from time import time

@contextmanager
def timer():
    s = time()
    yield
    e = time() - s
    print("{0}: {1} ms".format('Elapsed time', e))
```

函数使用方式如下：

```
import numpy as np

with timer():
    X = np.random.rand(1000)
```

输出执行这一命令行所消耗的时间如下：

Elapsed time: 0.0001399517059326172 ms

现在，你可以使用 scikit-learn 学习库的 **KneighborsClassifier** 检验训练和推理消耗的时间：

```
from sklearn.neighbors import KneighborsClassifier

# Defining properties of dataset
nT = 100000000 # Total number of values in our dataset
nF = 10 # Number of features
nE = int(nT / nF) # Number of examples

# Creating n x m matrix where n=100 and m=10
X = np.random.rand(nT).reshape(nE, nF)

# This will be a binary classification with labels 0 and 1
y = np.random.randint(2, size=nE)

# Data that we are going to score
scoring_data = np.random.rand(nF).reshape(1,-1)

# Create KNN classifier
knn = KNeighborsClassifier(11, algorithm='brute')

# Measure training time
with timer():
    knn.fit(X, y)

# Measure scoring time
with timer():
    knn.predict(scoring_data)
```

查看输出：

Elapsed time: 1.0800271034240723 ms
Elapsed time: 0.43231201171875 ms

为了与其他算法对比，可尝试一个其他分类器，如逻辑回归：

```
from sklearn.linear_model import LogisticRegression
log_res = LogisticRegression(C=1e5)

with timer():
    log_res.fit(X, y)

with timer():
    prediction = log_res.predict(scoring_data)
```

逻辑回归输出如下：

```
Elapsed time: 12.989803075790405 ms
Elapsed time: 0.00024318695068359375 ms
```

看起来差别很大！逻辑回归训练速度稍慢一些，但推理速度却快了很多。为什么这样呢？

以后会揭晓这个问题的答案。下面我们先简单了解 Python 代码分析，再深入讨论预测结果详情。

Python 代码分析
Code Profiling in Python

有些应用对机器学习模型的训练和推理有性能要求。比如，推荐引擎可能要求在 1 s 内生成推荐，如果延时超过 1 s，则要分析代码，了解产生延时的原因。代码分析能促进理解项目中各部分是如何执行的。分析的结果有指标，如调用数量，函数调用（包含/不包含子函数调用）的总时间，以及内存使用增量和内存使用总量。

Python 中的 cProfile 模块可查看每个函数的时间统计数据。这里有一个小例子：

```python
# cProfile
import cProfile

cProfile.run('np.std(np.random.rand(1000000))')
```

以上代码计算了从均匀分布中随机提取的 1000000 个样本的标准方差。输出显示了执行某个命令行的全部函数调用时间统计信息：

```
23 function calls in 0.025 seconds
   Ordered by: standard name
   ncalls tottime percall cumtime percall filename:lineno(function)
        1 0.001 0.001 0.025 0.025 <string>:1(<module>)
        1 0.000 0.000 0.007 0.007 _methods.py:133(_std)
        1 0.000 0.000 0.000 0.000 _methods.py:43(_count_reduce_items)
        1 0.006 0.006 0.007 0.007 _methods.py:86(_var)
        1 0.001 0.001 0.008 0.008 fromnumeric.py:2912(std)
        2 0.000 0.000 0.000 0.000 numeric.py:534(asanyarray)
        1 0.000 0.000 0.025 0.025 {built-in method builtins.exec}
        2 0.000 0.000 0.000 0.000 {built-in method builtins.hasattr}
```

```
        4 0.000 0.000 0.000 0.000 {built-in method builtins.isinstance}
        2 0.000 0.000 0.000 0.000 {built-in method builtins.issubclass}
        1 0.000 0.000 0.000 0.000 {built-in method builtins.max}
        2 0.000 0.000 0.000 0.000 {built-in method
numpy.core.multiarray.array}
        1 0.000 0.000 0.000 0.000 {method 'disable' of '_lsprof.Profiler'
objects}
        1 0.017 0.017 0.017 0.017 {method 'rand' of 'mtrand.RandomState'
objects}
        2 0.001 0.001 0.001 0.001 {method 'reduce' of 'numpy.ufunc'
objects}
```

0.025 秒内执行了 23 个函数调用，大部分时间消耗在随机数生成和标准方差计算上，与预期一致。

有一个很棒的库 snakeviz，用于 cProfile 可视化输出。构建一个名为 profiler_example_1.py 的文件，添加如下代码：

```
import numpy as np

np.std(np.random.rand(1000000))
```

在终端上，找出 profiler_example_1.py 所在的文件夹，运行以下命令：

```
python -m cProfile -o profiler_output -s cumulative profiler_example_1.py
```

此命令会创建一个称为 profiler_output 的文件，接下来可以使用 snakeviz 绘制图表。

性能统计数据可视化
Visualizing Performance Statistics

snakeviz 是基于浏览器的，支持按性能指标进行交互。snakeviz 使用 profiler 所生成的 profiler_output 文件创建可视化视图：

```
snakeviz profiler_output
```

此命令在本机 8080 端口跑一个小的 Web 服务器 127.0.0.1:8080，提供查看视图的地址，如 http://127.0.0.1:8080/snakeviz/.../2FAutomated_Machine_Learning%2FCh4_Automated_Algorithm_Selection%2Fprofiler_output。

打开地址，可看到一个旭日图（Sunburst），如图 4-2 所示，支持参数设置，如深度和临界值。

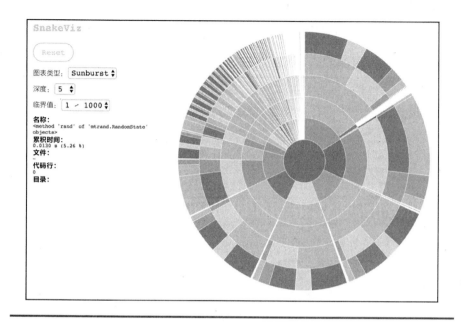

图 4-2　旭日图

鼠标悬浮到图表上，会显示函数名称、累积时间、文件、代码行和目录。你可以深入具体区域并查看详细信息。

若选择 Icicle 类型，效果如图 4-3 所示。

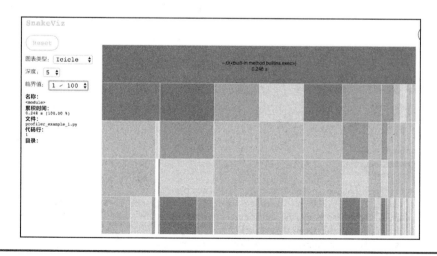

图 4-3　冰柱图

可调整图表类型、深度和临界值，查看哪些设置更合适。

滑动到底部，有一个相似的表格，如图 4-4 所示。

ncalls	tottime	percall	cumtime	percall	filename:lineno(function)
1	0.01301	0.01301	0.01301	0.01301	~:0(<method 'rand' of 'mtrand.RandomState' objects>)
19/18	0.09413	0.005229	0.09708	0.005393	~:0(<built-in method _imp.create_dynamic>)
1	0.004191	0.004191	0.005431	0.005431	_methods.py:86(_var)
127/1	0.003218	0.003218	0.2475	0.2475	~:0(<built-in method builtins.exec>)
1	0.000907	0.000907	0.2475	0.2475	profiler_example_1.py:1(<module>)
153/1	0.000865	0.000865	0.2282	0.2282	<frozen importlib._bootstrap>:958(_find_and_load)
141/1	0.000748	0.000748	0.228	0.228	<frozen importlib._bootstrap>:641(_load_unlocked)
153/1	0.000661	0.000661	0.2281	0.2281	<frozen importlib._bootstrap>:931(_find_and_load_unlocked)
2	0.001205	0.0006025	0.001205	0.0006025	~:0(<method 'reduce' of 'numpy.ufunc' objects>)
118/1	0.000446	0.000446	0.2279	0.2279	<frozen importlib._bootstrap_external>:672(exec_module)
1	0.00027	0.00027	0.01092	0.01092	random.py:38(<module>)
1	0.000261	0.000261	0.2273	0.2273	__init__.py:106(<module>)
118	0.02956	0.0002505	0.02956	0.0002505	~:0(<method 'read' of '_io.FileIO' objects>)
1	0.00023	0.00023	0.00023	0.00023	~:0(<function Random.seed at 0x105472510>)
1	0.000224	0.000224	0.000967	0.000967	getlimits.py:3(<module>)
1	0.000221	0.000221	0.001721	0.001721	core.py:21(<module>)
14	0.002977	0.0002126	0.002977	0.0002126	~:0(<built-in method posix.listdir>)

图 4-4　表格

如果根据 percall 列分类，就会发现 mtrand.RandomState 对象中的 rand 方法和 _var 方法是两个最耗时的调用。

使用这种方式可检验任何运行的命令，为理解和分析代码开一个好头。

从头开始实现 KNN
Implementing KNN from Scratch

我们已经见过KNN算法的应用，现在开始了解其实现。将如下代码块保存为 knn_prediction.py：

```python
import numpy as np
import operator

# distance module includes various distance functions
# You will use euclidean distance function to calculate distances between
scoring input and training dataset.
from scipy.spatial import distance
# Decorating function with @profile to get run statistics
@profile
def nearest_neighbors_prediction(x, data, labels, k):
```

```python
    # Euclidean distance will be calculated between example to be predicted
and examples in data
    distances = np.array([distance.euclidean(x, i) for i in data])

    label_count = {}
    for i in range(k):
        # Sorting distances starting from closest to our example
        label = labels[distances.argsort()[i]]
        label_count[label] = label_count.get(label, 0) + 1
    votes = sorted(label_count.items(), key=operator.itemgetter(1), reverse=True)

    # Return the majority vote
    return votes[0][0]

# Setting seed to make results reproducible
np.random.seed(23)

# Creating dataset, 20 x 5 matrix which means 20 examples with 5 features
for each
data = np.random.rand(100).reshape(20,5)

# Creating labels
labels = np.random.choice(2, 20)

# Scoring input
x = np.random.rand(5)

# Predicting class for scoring input with k=2
pred = nearest_neighbors_prediction(x, data, labels, k=2)
# Output is  '0'  in my case
```

分析这一函数，查看每一行代码需花费的时长。

逐行分析 Python 脚本
Profiling Your Python Script Line by Line

在终端上运行如下指令：

```
$ pip install line_profiler
```

安装完成后，可将以上代码片段保存为 knn_prediction.py 文件。

你可能已经注意到，下面定义了 nearest_neighbors_prediction：

```
@profile
def nearest_neighbors_prediction(x, data, labels, k):
    ...
```

这样，line_profiler 就能知道要分析哪个函数。运行以下命令，保存分析结果：

```
$ kernprof -l knn_prediction.py
```

输出如下：

```
Start
Wrote profile results to knn_prediction.py.lprof
```

查看分析器结果，代码如下：

```
$ python -m line_profiler knn_prediction.py.lprof
Timer unit: 1e-06 s

Total time: 0.001079 s
File: knn_prediction.py
Function: nearest_neighbors_prediction at line 24

Line #  Hits  Time  Per Hit  % Time  Line Contents
==============================================================
    24                                  @profile
    25                                  def nearest_neighbors_prediction(x, data, labels, k):
    26
    27                                      # Euclidean distance will be calculated between example to be predicted and examples in data
    28     1  1043.0  1043.0   96.7      distances = np.array([distance.euclidean(x, i) for i in data])
    29
    30     1     2.0     2.0    0.2      label_count = {}
    31     3     4.0     1.3    0.4      for i in range(k):
    32                                      # Sorting distances starting from closest to our example
    33     2    19.0     9.5    1.8          label = labels[distances.argsort()[i]]
    34     2     3.0     1.5    0.3          label_count[label] = label_count.get(label, 0) + 1
    35     1     8.0     8.0    0.7      votes = sorted(label_count.items(), key=operator.itemgetter(1), reverse=True)
    36
    37                                      # Return the majority vote
    38     1     0.0     0.0    0.0      return votes[0][0]
```

最耗时的部分是距离计算，在预期内。

 大 O 表示法中，KNN 算法的复杂度是 O(nm + kn)，其中 n 是样本的数量，m 是特征的数量，k 是算法的超参数。现在可以练习思考下其中的原因。

每个算法都会影响算法训练和评估时间的属性。要注意这些属性，对生产环境用例尤其重要。

4.4 线性与非线性
Linearity Versus Non-linearity

另一个考量因素是决策边界。一些算法，如逻辑回归或**支持向量机**（support vector machine，SVM），可学习到线性决策边界；而另一些算法，如树型算法，可学习到非线性决策边界。

虽然线性决策边界相对容易计算，也较易解释，但仍要注意线性算法在非线性关系中会产生的错误。

画出决策边界
Drawing Decision Boundaries

以下代码片段可检验不同类型算法的决策边界：

```
import matplotlib.cm as cm

# This function will scale training datatset and train given classifier.
# Based on predictions it will draw decision boundaries.

def draw_decision_boundary(clf, X, y, h = .01, figsize=(9,9),
boundary_cmap=cm.winter, points_cmap=cm.cool):

    # After you apply StandardScaler, feature means will be removed and all
    features will have unit variance.
    from sklearn.preprocessing import StandardScaler
    X = StandardScaler().fit_transform(X)

    # Splitting dataset to train and test sets.
    X_train, X_test, y_train, y_test = train_test_split(X, y, test_size=.4,
random_state=42)

    # Training given estimator on training dataset by invoking fit
function.
    clf.fit(X_train, y_train)

    # Each estimator has a score function.
    # Score will show you estimator's performance by showing metric
suitable to given classifier.
    # For example, for linear regression, it will output coefficient of
determination R^2 of the prediction.
    # For logistic regression, it will output mean accuracy.

    score = clf.score(X_test, y_test)
```

```python
    print("Score: %0.3f" % score)

    # Predict function of an estimator, will predict using trained model
    pred = clf.predict(X_test)

    # Figure is a high-level container that contains plot elements
    figure = plt.figure(figsize=figsize)

    # In current figure, subplot will create Axes based on given arguments
    (nrows, ncols, index)
    ax = plt.subplot(1, 1, 1)

    # Calculating min/max of axes
    x_min, x_max = X[:, 0].min() - 1, X[:, 0].max() + 1
    y_min, y_max = X[:, 1].min() - 1, X[:, 1].max() + 1

    # Meshgrid is usually used to evaluate function on grid.
    # It will allow you to create points to represent the space you operate
    xx, yy = np.meshgrid(np.arange(x_min, x_max, h), np.arange(y_min, y_max, h))

    # Generate predictions for all the point-pair created by meshgrid
    Z = clf.predict(np.c_[xx.ravel(), yy.ravel()])
    Z = Z.reshape(xx.shape)

    # This will draw boundary
    ax.contourf(xx, yy, Z, cmap=boundary_cmap)

    # Plotting training data
    ax.scatter(X_train[:, 0], X_train[:, 1], c=y_train, cmap=points_cmap, edgecolors='k')

    # Potting testing data
    ax.scatter(X_test[:, 0], X_test[:, 1], c=y_test, cmap=points_cmap, alpha=0.6, edgecolors='k')

    # Showing your masterpiece
    figure.show()
```

逻辑回归的决策边界
Decision Boundary of Logistic Regression

我们可以先使用逻辑回归测试此函数，代码如下：

```python
import numpy as np
import matplotlib.pyplot as plt
from matplotlib import cm
```

```python
# sklearn.linear_model includes regression models where target variable is
# a linear combination of input variables
from sklearn.linear_model import LogisticRegression

# make_moons is another useful function to generate sample data
from sklearn.datasets import make_moons
from sklearn.model_selection import train_test_split

X, y = make_moons(n_samples=1000, noise=0.1, random_state=0)

# Plot sample data
plt.scatter(X[:,0], X[:,1], c=y, cmap=cm.cool)
plt.show()
```

分布如图 4-5 所示。

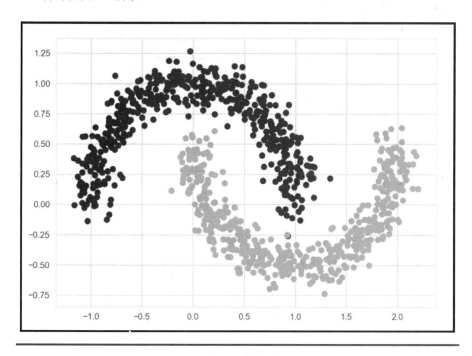

图 4-5　分布图

使用 draw_decision_boundary 函数将 LogisticRegression 的决策边界可视化：

```
draw_decision_boundary(LogisticRegression(), X, y)
```

决策边界分布图如图 4-6 所示。

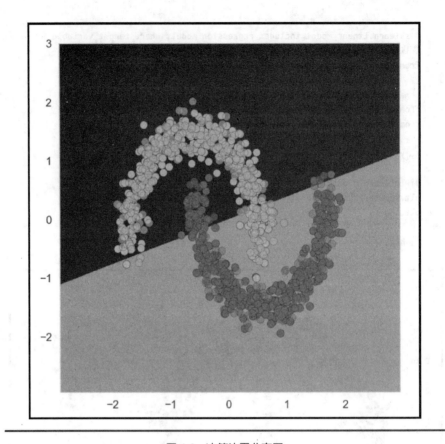

图 4-6 决策边界分布图

逻辑回归是广义线性模型之一，生成线性决策边界。线性决策边界是无法为上述数据集合理分类的。逻辑回归的输出是基于输入的权重之和计算出来的。由于输出不依赖于参数的积或商，产生的是线性决策边界。要解决这个问题，有许多方法，如正则化和特征映射，但也可采用其他非线性分类算法。

随机森林的决策边界
The Decision Boundary of Random Forest

随机森林是元估算器，它将构建许多不同的模型，集成所有模型的预测，汇总出最终的预测。随机森林可生成非线性决策边界，因为输入和输出之间没有线性关系。随机森林有许多超参数可设置，但为了简便，这里使用默认配置：

```
from sklearn.ensemble import RandomForestClassifier

draw_decision_boundary(RandomForestClassifier(), X, y)
```

以上代码得到随机森林决策边界分布图如图 4-7 所示。

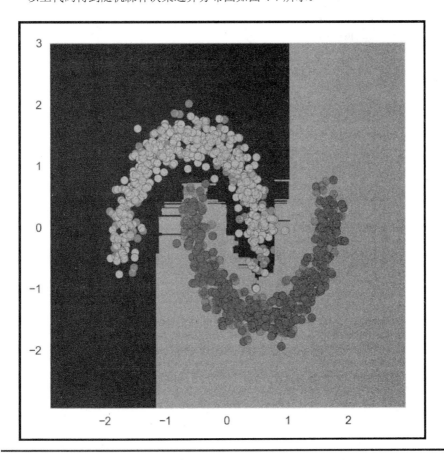

图 4-7　随机森林决策边界分布图

看起来不错！每个算法会根据自身的工作机制提供不同的决策边界，建议你分别试验这些估算器，更好地理解这些算法的行为。

常用机器学习算法
Commonly Used Machine Learning Algorithms

下面列出了常用的监督算法和无监督算法，推荐练习。其中大多数算法都可以在 scikit-learn 中找到。

- 监督算法为：
  - 线性回归；
  - 逻辑回归；
  - KNN；
  - 随机森林；
  - 提升算法（GBM、XGBoost 和 LightGBM）；
  - SVM；
  - 神经网络。
- 无监督算法：
  - K-means；
  - 层次聚类；
  - 主成分分析；
  - 混合模型；
  - 自编码。

4.5 必要特征转换
Necessary Feature Transformations

你可能留意到前面章节中训练机器学习算法的特征是经过缩放了的。特征转换一般是保证机器学习算法正常生效的必要条件。比如，采用正则化的机器学习算法的首要原则是对特征进行归一化处理。以下用例中，在为机器学习算法准备数据集时，需要进行特征转换。

- SVM 的预期输入是标准范围内的数据。变量要先归一化处理后，再输入算法中。
- 主成分分析（PCA）根据方差最大化理论将特征映射到另一空间，选择出数据集中方差最大的成分，排除其他成分，实现降维。采用 PCA 时，要对特征归一化处理，因为有些特征因规格过大而几乎占据全部方差。归一化处理可消除规格上的差异，下一节会展示一些示例。

- 高维数据集经常要采用正则化回归，应该把变量归一化，控制规模差异，因为正则化还是受规模变化影响的。
- 朴素贝叶斯算法中，特征列和标签列应该是类别型数据，可采用分箱法将连续变量离散化。
- 在时间序列中，可对指数级上涨趋势进行对数转换，得到线性趋势和常量方差。
- 处理非数值型变量，如类别型数据，可利用独热编码、伪编码或特征哈希将类别型特征编码成数值型特征。

4.6 监督机器学习
Supervised ML

除上一节提到的特征转换之外，每种机器学习算法都有超参数空间需要优化。发现最佳机器学习流水线的过程可以理解成遍历所有配置空间，智能尝试各种选项，直到找到性能最好的流水线。

auto-sklearn 是实现这一目标非常有用的库，在简介一章中介绍的示例已经体现出此库的简单易用性。本节将阐释此库成功实施背后的深层原理。

auto-sklearn 采用元学习，根据已有数据集的特性，选择有潜力的数据/特征处理器和机器学习算法。了解预处理器、分类器和回归器，请参见如下链接。

- 分类器（https://github.com/automl/auto-sklearn/tree/master/autosklearn/pipeline/components/classification）。
- 回归器（https://github.com/automl/auto-sklearn/tree/master/autosklearn/pipeline/ components/regression）。
- 预处理器（https://github.com/automl/auto-sklearn/tree/master/autosklearn/pipeline/components/feature_preprocessing）。

元学习模拟数据科学家的体验,在不同的数据集中分析机器学习流水线的性能,将发现的结果匹配到新数据集中,从而推荐初始配置。

元学习形成初始配置后,贝叶斯优化负责各流水线的超参数优化,排名高的流水线用于创建一套集成法,性能可能超过任何成员,也利于避免过度拟合。

auto-sklearn 默认配置
Default Configuration of auto-sklearn

接下来即将创建 AutoSklearnClassifier 对象,要注意默认配置。运行以下代码可看到默认配置:

```
from autosklearn.classification import AutoSklearnClassifier
AutoSklearnClassifier?
```

在 Python 中,函数后加? 可输出很多有用信息,如签名、文件字符串、参数说明、属性和文件位置。

签名中有默认值,代码如下:

```
Init signature: AutoSklearnClassifier(time_left_for_this_task=3600,
per_run_time_limit=360, initial_configurations_via_metalearning=25,
ensemble_size=50, ensemble_nbest=50, seed=1, ml_memory_limit=3072,
include_estimators=None, exclude_estimators=None,
include_preprocessors=None, exclude_preprocessors=None,
resampling_strategy='holdout', resampling_strategy_arguments=None,
tmp_folder=None, output_folder=None,
delete_tmp_folder_after_terminate=True,
delete_output_folder_after_terminate=True, shared_mode=False,
disable_evaluator_output=False, get_smac_object_callback=None,
smac_scenario_args=None)
```

例如,time_left_for_this_task 值默认为 60 分钟。若处理很复杂的数据集,应将此参数值增大,提高发现更优机器学习流水线的概率。

另一个 per_run_time_limit 值默认为 6 分钟。许多机器学习算法的训练时间与输入数据的大小成比例,而且受算法的复杂度影响。此参数应酌情设置。

ensemble_size 和 ensemble_nbest 是集成相关参数,用于设置集成中要包含的最佳模型的大小和数量。

ml_memory_limit 是一个重要参数,因为如果算法要求的内存超过此限制,则训练就会被取消。

以下参数可用于在机器学习流水线中包含/排除某些数据预处理器或估算器：

include_estimators、exclude_estimators、include_preprocessors 及 exclude_preprocessors。

resampling_strategy 提供处理过度拟合的方法。

查看签名中的其他参数，看是否有需要根据环境调整某些具体的参数。

找出产品线预测的最佳机器学习流水线
Finding the Best ML Pipeline for Product Line Prediction

首先编写一个包装器函数，编码类别型变量，准备数据集：

```
# Importing necessary variables
import numpy as np
import pandas as pd
from autosklearn.classification import AutoSklearnClassifier
from autosklearn.regression import AutoSklearnRegressor
from sklearn.model_selection import train_test_split
from sklearn.metrics import accuracy_score
from sklearn.preprocessing import LabelEncoder
import wget
import pandas as pd

# Machine learning algorithms work with numerical inputs and you need to
transform all non-numerical inputs to numerical ones
# Following snippet encode the categorical variables

link_to_data =
'https://apsportal.ibm.com/exchange-api/v1/entries/8044492073eb964f46597b4b
e06ff5ea/data?accessKey=9561295fa407698694b1e254d0099600'
filename = wget.download(link_to_data)

print(filename)
# GoSales_Tx_NaiveBayes.csv

df = pd.read_csv('GoSales_Tx_NaiveBayes.csv')
df.head()
```

输出 DataFrame 的前 5 条记录：

```
# PRODUCT_LINE GENDER AGE MARITAL_STATUS PROFESSION
# 0 Personal Accessories M 27 Single Professional
# 1 Personal Accessories F 39 Married Other
# 2 Mountaineering Equipment F 39 Married Other
# 3 Personal Accessories F 56 Unspecified Hospitality
# 4 Golf Equipment M 45 Married Retired
```

这个数据集中有 4 个特征（GENDER、AGE、MARITAL_STATUS 和 PROFESSION）和一个标签（PRODUCT_LINE）。目标是预测客户感兴趣的产品线。

使用 LabelEncoder 将特征和标签文本数据进行编码：

```
df = df.apply(LabelEncoder().fit_transform)
df.head()
```

编码 label 列：

```
#   PRODUCT_LINE GENDER AGE MARITAL_STATUS PROFESSION
# 0  4            1      27  1              3
# 1  4            0      39  0              2
# 2  2            0      39  0              2
# 3  4            0      56  2              1
# 4  1            1      45  0              5
```

如上所示，全部类别列均被编码。要记住，在 auto-sklearn 的 API 中，有一个 feat_type 参数，可将列定义为类别型或数值型：

feat_type : list, optional (default=None)

len(X.shape[1]) 中的 str 列表是描述属性类型的。可选类型有类别型和数值型。类别型属性会被自动执行独热编码。类别型属性中编码的值必须是整型，如从 sklearn.proprocessing.LabelEncoder 中获得的整型。

但是，本例中也可使用 pandas DataFrame 中的 apply 函数。

以下包装器函数会采用 auto-sklearn 中的 auto-classification 或 auto-regression 算法处理输入数据，运行实验：

```
# Function below will encode the target variable if needed
def encode_target_variable(df=None, target_column=None, y=None):

    # Below section will encode target variable if given data is pandas dataframe
    if df is not None:
        df_type = isinstance(df, pd.core.frame.DataFrame)

        # Splitting dataset as train and test data sets
        if df_type:

            # If column data type is not numeric then labels are encoded
            if not np.issubdtype(df[target_column].dtype, np.number):
                le = preprocessing.LabelEncoder()
                df[target_column] = le.fit_transform(df[target_column])
            return df[target_column]
```

```
            return df[target_column]
    # Below section will encode numpy array.
    else:
    
        # numpy array's data type is not numeric then labels are encoded
        if not np.issubdtype(y.dtype, np.number):
            le = preprocessing.LabelEncoder()
            y = le.fit_transform(y)
            return y
        
        return y


# Create a wrapper function where you can specify the type of learning
problem
def supervised_learner(type, X_train, y_train, X_test, y_test):

    if type == 'regression':
        # You can play with time related arguments for discovering more
pipelines
        automl = AutoSklearnRegressor(time_left_for_this_task=7200,
per_run_time_limit=720)
    else:
        automl = AutoSklearnClassifier(time_left_for_this_task=7200,
per_run_time_limit=720)

    # Training estimator based on learner type
    automl.fit(X_train, y_train)

    # Predicting labels on test data
    y_hat = automl.predict(X_test)

    # Calculating accuracy_score
    metric = accuracy_score(y_test, y_hat)

# Return model, labels and metric
return automl, y_hat, metric

# In function below, you need to provide numpy array or pandas dataframe
together with the name of the target column as arguments
def supervised_automl(data, target_column=None, type=None, y=None):

    # First thing is to check whether data is pandas dataframe
    df_type = isinstance(data, pd.core.frame.DataFrame)

    # Based on data type, you will split dataset as train and test data
sets
    if df_type:
        # This is where encode_target_variable function is used before data
split
        data[target_column] = encode_target_variable(data, target_column)
        X_train, X_test, y_train, y_test = \
            train_test_split(data.loc[:, data.columns != target_column],
```

```
                    data[target_column], random_state=1)
            else:
                y_encoded = encode_target_variable(y=y)
                X_train, X_test, y_train, y_test = train_test_split(X, y_encoded,
    random_state=1)

        # If learner type is given, then you invoke supervied_learner
        if type != None:
            automl, y_hat, metric = supervised_learner(type, X_train, y_train,
    X_test, y_test)

        # If type of learning problem is not given, you need to infer it
        # If there are more than 10 unique numerical values, problem will be
    treated as regression problem,
        # Otherwise, classification problem

        elif len(df[target_column].unique()) > 10:
                print("""There are more than 10 uniques numerical values in
    target column.
                Treating it as regression problem.""")
                automl, y_hat, metric = supervised_learner('regression',
    X_train, y_train, X_test, y_test)
        else:
                automl, y_hat, metric = supervised_learner('classification',
    X_train, y_train, X_test, y_test)

        # Return model, labels and metric
        return automl, y_hat, metric
```

运行代码，查看结果：

```
automl, y_hat, metric = supervised_automl(df, target_column='PRODUCT_LINE')
```

以下输出展示了所选的模型及参数：

```
automl.get_models_with_weights()
[(1.0,
SimpleClassificationPipeline({'balancing:strategy': 'none',
'categorical_encoding:__choice__': 'no_encoding', 'classifier:__choice__':
'gradient_boosting', 'imputation:strategy': 'most_frequent',
'preprocessor:__choice__': 'feature_agglomeration', 'rescaling:__choice__':
'robust_scaler', 'classifier:gradient_boosting:criterion': 'friedman_mse',
'classifier:gradient_boosting:learning_rate': 0.6019977814828193,
'classifier:gradient_boosting:loss': 'deviance',
'classifier:gradient_boosting:max_depth': 5,
'classifier:gradient_boosting:max_features': 0.4884281825655421,
'classifier:gradient_boosting:max_leaf_nodes': 'None',
'classifier:gradient_boosting:min_impurity_decrease': 0.0,
'classifier:gradient_boosting:min_samples_leaf': 20,
'classifier:gradient_boosting:min_samples_split': 7,
'classifier:gradient_boosting:min_weight_fraction_leaf': 0.0,
'classifier:gradient_boosting:n_estimators': 313,
'classifier:gradient_boosting:subsample': 0.3242201709371377,
'preprocessor:feature_agglomeration:affinity': 'cosine',
```

```
'preprocessor:feature_agglomeration:linkage': 'complete',
'preprocessor:feature_agglomeration:n_clusters': 383,
'preprocessor:feature_agglomeration:pooling_func': 'mean',
'rescaling:robust_scaler:q_max': 0.75, 'rescaling:robust_scaler:q_min':
0.25},
  dataset_properties={
    'task': 1,
    'sparse': False,
    'multilabel': False,
    'multiclass': False,
    'target_type': 'classification',
    'signed': False}))]
```

根据经验会发现所选的是梯度提升算法，符合预期。因为在当前机器学习领域中，提升算法是最先进的，最常用的有 XGBoost、LightGBM 和 CatBoost。

auto-sklearn 支持 sklearn 的 `GradientBoostingClassifier`。XGBoost 属于集成性算法，目前不可用，但预计很快可以回归。

找出网络异常检测的最佳机器学习流水线
Finding the Best Machine Learning Pipeline for Network Anomaly Detection

在机器学习领域另一个常用的数据集上运行这个流水线。KDDCUP 99 数据集是 1998 DARPA Intrusion Detection System Evaluation 数据集的 tcpdump 部分，目标是检测网络入侵。此数据集中包括数值型特征，因为这样设置 AutoML 流水线更容易。

```
# You can import this dataset directly from sklearn
from sklearn.datasets import fetch_kddcup99

# Downloading subset of whole dataset
dataset = fetch_kddcup99(subset='http', shuffle=True, percent10=True)
# Downloading https://ndownloader.figshare.com/files/5976042
# [INFO] [17:43:19:sklearn.datasets.kddcup99] Downloading
https://ndownloader.figshare.com/files/5976042

X = dataset.data
y = dataset.target

# 58725 examples with 3 features
X.shape
# (58725, 3)

y.shape
```

```
(58725,)

# 5 different classes to represent network anomaly
from pprint import pprint
pprint(np.unique(y))
# array([b'back.', b'ipsweep.', b'normal.', b'phf.', b'satan.'],
dtype=object)

automl, y_hat, metric = supervised_automl(X, y=y, type='classification')
```

4.7 无监督 AutoML
Unsupervised AutoML

数据集没有目标变量时，可根据不同的特性用聚类算法去探索。这些算法将示例进行分组，每个组内的示例都尽量接近，而不同组的示例则尽量不同。

由于进行分析时基本没有标签，所以有一个性能指标可检验算法所进行的分割质量如何。

这个指标称为**轮廓系数**（silhouette coefficient）。轮廓系数有助于理解两点。

- **黏合度**（cohesion）：聚类之间的相似性。
- **分离度**（separation）：聚类之间的差异性。

取值在 1 和 −1 之间，越接近 1，表示聚类形态越好。

若训练数据有标签，也可使用其他指标，如同质性和完整性，后面的章节会提到。

聚类算法可完成许多不同的任务，如查找相似的用户、歌曲或图片，检测模式中的关键趋势和变化，理解社交网络中的社群结构。

常用聚类算法
Commonly Used Clustering Algorithms

常用聚类算法有两种：距离算法和概率算法。比如，k-means 和 DBSCAN（**具有噪声的基于密度的聚类算法**）是距离算法，而高斯混合模型是概率算法。

距离算法有各种距离度量方法可用，常用欧氏距离指标。

概率算法假定存在一个包含未知参数的概率分布生成过程，其目标是从数据中计算这些参数。

聚类算法有许多，应依据数据特性选择合适的算法。例如，k-means 适用于有形心的聚类，要求数据的聚类大小均匀，且形状是中间凸出的。这意味着 k-means 不适合细长聚类或非规则形状的流形。当数据中的聚类大小不均或形状呈凸形时，建议采用 DBSCAN 这种适用于任意形状聚类区的算法。

对数据了解一二之后，会更容易找出合适的算法，但如果不了解数据呢？许多时候探索分析时，可能很难立即对数据理出头绪。遇到这种情况时，可试试自动化无监督机器学习流水线，有助于更好地理解数据特性。

但是，做这类分析时还是要谨慎。因为后续运作会受到分析结果的驱动，若稍不留意，可能就误入歧途了。

使用 sklearn 创建样本数据集
Creating Sample Datasets with Sklearn

在 sklearn 中，有些有用的方法可创建测试算法用的样本数据集：

```
# Importing necessary libraries for visualization
import matplotlib.pyplot as plt
import seaborn as sns

# Set context helps you to adjust things like label size, lines and various elements
# Try "notebook", "talk" or "paper" instead of "poster" to see how it changes
sns.set_context('poster')

# set_color_codes will affect how colors such as 'r', 'b', 'g' will be interpreted
sns.set_color_codes()

# Plot keyword arguments will allow you to set things like size or line width to be used in charts.
plot_kwargs = {'s': 10, 'linewidths': 0.1}

import numpy as np
import pandas as pd

# Pprint will better output your variables in console for readability
from pprint import pprint
```

```python
# Creating sample dataset using sklearn samples_generator
from sklearn.datasets.samples_generator import make_blobs
from sklearn.preprocessing import StandardScaler

# Make blobs will generate isotropic Gaussian blobs
# You can play with arguments like center of blobs, cluster standard
deviation
centers = [[2, 1], [-1.5, -1], [1, -1], [-2, 2]]
cluster_std = [0.1, 0.1, 0.1, 0.1]

# Sample data will help you to see your algorithms behavior
X, y = make_blobs(n_samples=1000,
                  centers=centers,
                  cluster_std=cluster_std,
                  random_state=53)

# Plot generated sample data
plt.scatter(X[:, 0], X[:, 1], **plot_kwargs)
plt.show()
```

以上代码输出如图 4-8 所示。

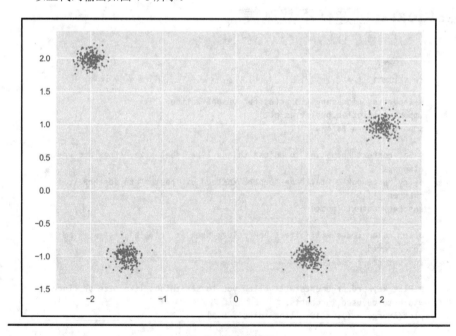

图 4-8　样本数据图

cluster_std 影响离散量。改成[0.4, 0.5, 0.6, 0.5]，再试一次：

```
cluster_std = [0.4, 0.5, 0.6, 0.5]
```

```
X, y = make_blobs(n_samples=1000,
                  centers=centers,
                  cluster_std=cluster_std,
                  random_state=53)

plt.scatter(X[:, 0], X[:, 1], **plot_kwargs)
plt.show()
```

以上代码输出如图 4-9 所示。

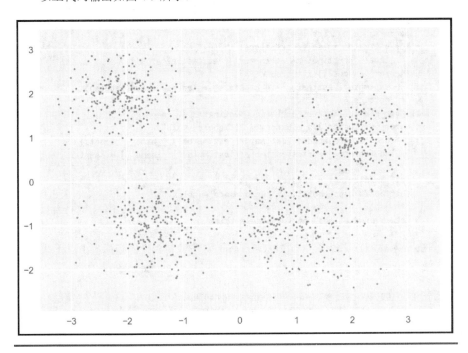

图 4-9　离散后的样本数据图

这样看起来更真实可信了！

现在编写一个小 Python 程序类，使用这些好方法创建无监督学习实验。首先，使用 fit_predict 方法将一个或多个聚类算法拟合到样本数据集中：

```
class Unsupervised_AutoML:

    def __init__(self, estimators=None, transformers=None):
        self.estimators = estimators
        self.transformers = transformers
        pass
```

Unsupervised_AutoML 类会初始化为一套估算器和转换器。第二个类方法是 fit_predict：

```python
def fit_predict(self, X, y=None):
    """
    fit_predict will train given estimator(s) and predict cluster
membership for each sample
    """

    # This dictionary will hold predictions for each estimator
    predictions = []
    performance_metrics = {}
    for estimator in self.estimators:
        labels = estimator['estimator'](*estimator['args'],
**estimator['kwargs']).fit_predict(X)
        estimator['estimator'].n_clusters_ = len(np.unique(labels))
        metrics = 
self._get_cluster_metrics(estimator['estimator'].__name__,
estimator['estimator'].n_clusters_, X, labels, y)
        predictions.append({estimator['estimator'].__name__: labels})
        performance_metrics[estimator['estimator'].__name__] = metrics

    self.predictions = predictions
    self.performance_metrics = performance_metrics

    return predictions, performance_metrics
```

fit_predict 方法使用 _get_cluster_metrics 方法得到性能指标，由以下代码定义：

```python
# Printing cluster metrics for given arguments
def _get_cluster_metrics(self, name, n_clusters_, X, pred_labels,
true_labels=None):
    from sklearn.metrics import homogeneity_score, \
        completeness_score, \
        v_measure_score, \
        adjusted_rand_score, \
        adjusted_mutual_info_score, \
        silhouette_score

    print("""################# %s metrics ##################""" % name)
    if len(np.unique(pred_labels)) >= 2:

        silh_co = silhouette_score(X, pred_labels)

        if true_labels is not None:

            h_score = homogeneity_score(true_labels, pred_labels)
            c_score = completeness_score(true_labels, pred_labels)
            vm_score = v_measure_score(true_labels, pred_labels)
            adj_r_score = adjusted_rand_score(true_labels, pred_labels)
            adj_mut_info_score = adjusted_mutual_info_score(true_labels,
```

```
        pred_labels)
            metrics = {"Silhouette Coefficient": silh_co,
                       "Estimated number of clusters": n_clusters_,
                       "Homogeneity": h_score,
                       "Completeness": c_score,
                       "V-measure": vm_score,
                       "Adjusted Rand Index": adj_r_score,
                       "Adjusted Mutual Information": adj_mut_info_score}

            for k, v in metrics.items():
                print("\t%s: %0.3f" % (k, v))

            return metrics

        metrics = {"Silhouette Coefficient": silh_co,
                   "Estimated number of clusters": n_clusters_}

        for k, v in metrics.items():
            print("\t%s: %0.3f" % (k, v))

        return metrics

    else:
        print("\t# of predicted labels is {}, can not produce metrics.
\n".format(np.unique(pred_labels)))
```

`_get_cluster_metrics` 方法计算指标,如 `homogeneity_score`(同质评分)、`completeness_score`(完整性评分)、`v_measure_score`(v 指标评分)、`adjusted_rand_score`(调整随机评分)、`adjusted_mutual_info_sco`(校正后互信息评分)和 `silhouette_score`(轮廓系数)。这些指标有利于评估聚类是否分割良好,且可衡量聚类内和聚类间的相似性。

k-means 算法实践
k-means Algorithm in Action

应用 k-means 算法,查看效果:

```
from sklearn.cluster import KMeans

estimators = [{'estimator': KMeans, 'args':(), 'kwargs':{'n_clusters': 4}}]

unsupervised_learner = Unsupervised_AutoML(estimators)
```

可查看 estimators:

```
unsupervised_learner.estimators
```

输出如下：

```
[{'args': (),
  'estimator': sklearn.cluster.k_means_.KMeans,
  'kwargs': {'n_clusters': 4}}]
```

可调用 fit_predict，获得 predictions 和 performanace_metrics，代码如下：

```
predictions, performance_metrics = unsupervised_learner.fit_predict(X, y)
```

指标会输出到控制台：

```
################ KMeans metrics ##################
    Silhouette Coefficient: 0.631
    Estimated number of clusters: 4.000
    Homogeneity: 0.951
    Completeness: 0.951
    V-measure: 0.951
    Adjusted Rand Index: 0.966
    Adjusted Mutual Information: 0.950
```

以后也可随时打印：

```
pprint(performance_metrics)
```

输出估算器名称及其指标：

```
{'KMeans': {'Silhouette Coefficient': 0.9280431207593165, 'Estimated number
of clusters': 4, 'Homogeneity': 1.0, 'Completeness': 1.0, 'V-measure': 1.0,
'Adjusted Rand Index': 1.0, 'Adjusted Mutual Information': 1.0}}
```

加入另一个类方法，绘制已知估算器和预测标签的聚类图：

```python
# plot_clusters will visualize the clusters given predicted labels
def plot_clusters(self, estimator, X, labels, plot_kwargs):

    palette = sns.color_palette('deep', np.unique(labels).max() + 1)
    colors = [palette[x] if x >= 0 else (0.0, 0.0, 0.0) for x in labels]

    plt.scatter(X[:, 0], X[:, 1], c=colors, **plot_kwargs)
    plt.title('{} Clusters'.format(str(estimator.__name__)), fontsize=14)
    plt.show()
```

看看用法：

```
plot_kwargs = {'s': 12, 'linewidths': 0.1}
unsupervised_learner.plot_clusters(KMeans,
                                   X,
                                   unsupervised_learner.predictions[0]['KMeans'],
                                   plot_kwargs)
```

上述代码输出如图 4-10 所示。

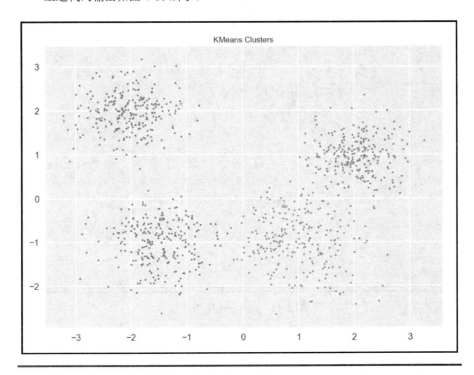

图 4-10　k-means 聚类图 a

本例中，聚类大小均匀，彼此界限清晰，但探索分析时，仍应尝试不同的超参数以检验效果。

本章后面部分会带你编写一个包装器函数，应用一组聚类算法和超参数进行结果检验。现在再看一个 k-means 算法用得不太好的例子。

数据集有不同的统计属性，比如方差差异，k-means 算法无法正确地识别出聚类：

```
X, y = make_blobs(n_samples=2000, centers=5, cluster_std=[1.7, 0.6, 0.8,
1.0, 1.2], random_state=220)

# Plot sample data
plt.scatter(X[:, 0], X[:, 1], **plot_kwargs)
plt.show()
```

上述代码输出如图 4-11 所示。

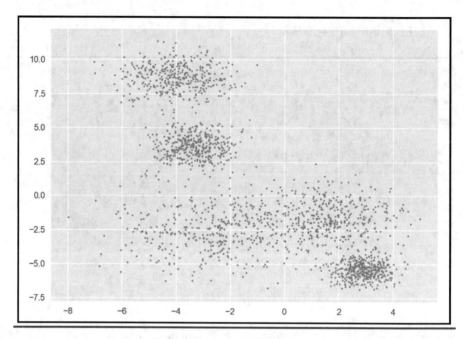

图 4-11　k-means 聚类图 b

虽然样本数据集有 5 个中心，但不是很明显，也可能是 4 个聚类：

```
from sklearn.cluster import KMeans

estimators = [{'estimator': KMeans, 'args':(), 'kwargs':{'n_clusters': 4}}]

unsupervised_learner = Unsupervised_AutoML(estimators)

predictions, performance_metrics = unsupervised_learner.fit_predict(X, y)
```

控制台中显示指标如下：

```
################ KMeans metrics ##################
    Silhouette Coefficient: 0.549
    Estimated number of clusters: 4.000
    Homogeneity: 0.729
    Completeness: 0.873
    V-measure: 0.795
    Adjusted Rand Index: 0.702
    Adjusted Mutual Information: 0.729
```

k-means 聚类分布绘图代码如下：

```
plot_kwargs = {'s': 12, 'linewidths': 0.1}
```

```
unsupervised_learner.plot_clusters(KMeans,
                                    X,
                  unsupervised_learner.predictions[0]['KMeans'],
                                    plot_kwargs)
```

上述代码输出如图 4-12 所示。

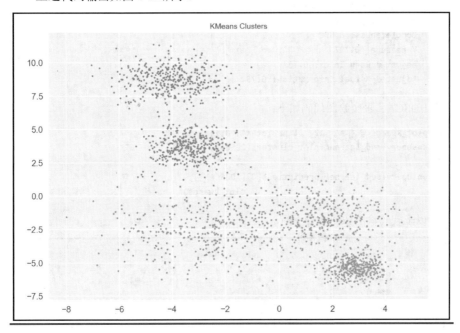

图 4-12　k-means 聚类图 c

本例中，红色（深灰）和底部绿色（浅灰）之间的点看起来形成了一个大聚类。k-means 是基于形心周边点的平均值计算形心的。这时就需要调整方法了。

DBSCAN 算法实践
The DBSCAN Algorithm in Action

DBSCAN 是处理非平面几何形状和不均匀聚类大小的聚类算法之一。看看 DBSCAN 的用法，代码如下：

```
from sklearn.cluster import DBSCAN

estimators = [{'estimator': DBSCAN, 'args':(), 'kwargs':{'eps': 0.5}}]

unsupervised_learner = Unsupervised_AutoML(estimators)
```

```
predictions, performance_metrics = unsupervised_learner.fit_predict(X, y)
```

控制台中输出指标如下:

```
################# DBSCAN metrics ###################
  Silhouette Coefficient: 0.231
  Estimated number of clusters: 12.000
  Homogeneity: 0.794
  Completeness: 0.800
  V-measure: 0.797
  Adjusted Rand Index: 0.737
  Adjusted Mutual Information: 0.792
```

BDSCAN 聚类绘图代码如下:

```
plot_kwargs = {'s': 12, 'linewidths': 0.1}
unsupervised_learner.plot_clusters(DBSCAN,
                                   X,
unsupervised_learner.predictions[0]['DBSCAN'],
                                   plot_kwargs)
```

以上代码输出如图 4-13 所示。

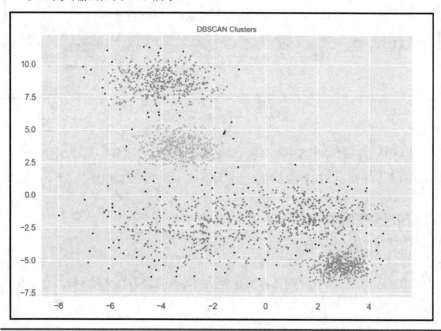

图 4-13　DBSCAN 聚类图

在 k-means 案例中的红色（深灰）和底部绿色（浅灰）聚类之间的边界似乎消

失了，但有意思的是，出现了一些小的聚类，有些点按距离算不属于任何一个大的聚类。

DBSCAN 有一个 esp(epsilon)超参数，与同一邻近区内点之间的接近度相关。尝试调整此参数，看看算法表现如何。

对了解较少的数据做探索分析时，视觉线索很重要，因为指标可能有误导性，不是每个聚类算法都能用相似指标进行评价的。

凝聚聚类算法实践
Agglomerative Clustering Algorithm in Action

最后尝试凝聚聚类算法，代码如下：

```python
from sklearn.cluster import AgglomerativeClustering

estimators = [{'estimator': AgglomerativeClustering, 'args':(),
'kwargs':{'n_clusters': 4, 'linkage': 'ward'}}]
unsupervised_learner = Unsupervised_AutoML(estimators)

predictions, performance_metrics = unsupervised_learner.fit_predict(X, y)
```

控制台输出指标如下：

```
################ AgglomerativeClustering metrics ###################
  Silhouette Coefficient: 0.546
  Estimated number of clusters: 4.000
  Homogeneity: 0.751
  Completeness: 0.905
  V-measure: 0.820
  Adjusted Rand Index: 0.719
  Adjusted Mutual Information: 0.750
```

AgglomerativeClusterig 聚类绘制代码如下：

```python
plot_kwargs = {'s': 12, 'linewidths': 0.1}
unsupervised_learner.plot_clusters(AgglomerativeClustering,
                                   X,
unsupervised_learner.predictions[0]['AgglomerativeClustering'],
                                   plot_kwargs)
```

以上代码输出如图 4-14 所示。

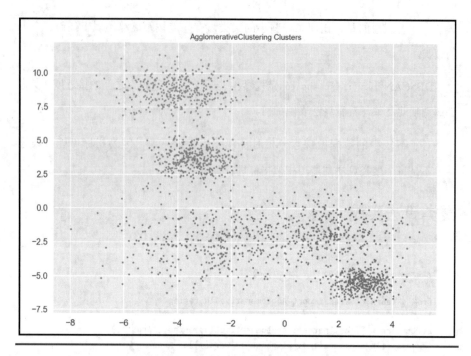

图 4-14　AgglomerativeClustering 聚类图

AgglomerativeClustering 在本例中的表现与 k-means 的相似，区别较小。

无监督学习的简单自动化
Simple Automation of Unsupervised Learning

你应将整个发现过程自动化，在不同的超参数设置中尝试不同的算法。以下代码展示了一种简单的自动化方法：

```
# You will create a list of algorithms to test
from sklearn.cluster import MeanShift, estimate_bandwidth, SpectralClustering
from hdbscan import HDBSCAN

# bandwidth estimate for MeanShift algorithm to work properly
bandwidth = estimate_bandwidth(X, quantile=0.3, n_samples=100)

estimators = [{'estimator': KMeans, 'args': (), 'kwargs': {'n_clusters': 5}},
              {'estimator': DBSCAN, 'args': (), 'kwargs': {'eps': 0.5}},
              {'estimator': AgglomerativeClustering, 'args': (),
```

```
                    'kwargs': {'n_clusters': 5, 'linkage': 'ward'}},
                                {'estimator': MeanShift, 'args': (), 'kwargs':
{'cluster_all': False, "bandwidth": bandwidth, "bin_seeding": True}},
                                {'estimator': SpectralClustering, 'args': (),
'kwargs': {'n_clusters':5}},
                                {'estimator': HDBSCAN, 'args': (), 'kwargs':
{'min_cluster_size':15}}]


unsupervised_learner = Unsupervised_AutoML(estimators)

predictions, performance_metrics = unsupervised_learner.fit_predict(X, y)
```

控制台中可见如下指标:

```
################ KMeans metrics ##################
  Silhouette Coefficient: 0.592
  Estimated number of clusters: 5.000
  Homogeneity: 0.881
  Completeness: 0.882
  V-measure: 0.882
  Adjusted Rand Index: 0.886
  Adjusted Mutual Information: 0.881


################ DBSCAN metrics ##################
  Silhouette Coefficient: 0.417
  Estimated number of clusters: 5.000
  ...
################ AgglomerativeClustering metrics ##################
  Silhouette Coefficient: 0.581
  Estimated number of clusters: 5.000
  ...
################ MeanShift metrics ##################
  Silhouette Coefficient: 0.472
  Estimated number of clusters: 3.000
  ...
################ SpectralClustering metrics ##################
  Silhouette Coefficient: 0.420
  Estimated number of clusters: 5.000
  ...
################ HDBSCAN metrics ##################
  Silhouette Coefficient: 0.468
  Estimated number of clusters: 6.000
  ...
```

稍后可以打印标签和指标,每个算法都有一个标签和一些指标:

```
pprint(predictions)
[{'KMeans': array([3, 1, 4, ..., 0, 1, 2], dtype=int32)},
```

```
 {'DBSCAN': array([ 0, 0, 0, ..., 2, -1, 1])},
 {'AgglomerativeClustering': array([2, 4, 0, ..., 3, 2, 1])},
 {'MeanShift': array([0, 0, 0, ..., 1, 0, 1])},
 {'SpectralClustering': array([4, 2, 1, ..., 0, 1, 3], dtype=int32)},
 {'HDBSCAN': array([ 4, 2, 3, ..., 1, -1, 0])}]

pprint(performance_metrics)
{'AgglomerativeClustering': {'Adjusted Mutual Information':
0.8989601162598674,
                             'Adjusted Rand Index': 0.9074196173180163,
                             ...},
 'DBSCAN': {'Adjusted Mutual Information': 0.5694008711591612,
            'Adjusted Rand Index': 0.4685215791890368,
            ...},
 'HDBSCAN': {'Adjusted Mutual Information': 0.7857262723310214,
             'Adjusted Rand Index': 0.7907512089039799,
             ...},
 'KMeans': {'Adjusted Mutual Information': 0.8806038790635883,
            'Adjusted Rand Index': 0.8862210038915361,
            ...},
 'MeanShift': {'Adjusted Mutual Information': 0.45701704058584275,
               'Adjusted Rand Index': 0.4043364504640998,
               ...},
 'SpectralClustering': {'Adjusted Mutual Information': 0.7628653432724043,
                        'Adjusted Rand Index': 0.7111907598912597,
                        ...}}
```

可使用 plot_clusters 方法同样将预测结果可视化展现出来。再编写一个类方法，将这个实验中用到的全部估算器的聚类都绘制出来，代码如下：

```
def plot_all_clusters(self, estimators, labels, X, plot_kwargs):

    fig = plt.figure()

    for i, algorithm in enumerate(labels):

        quotinent = np.divide(len(estimators), 2)

        # Simple logic to decide row and column size of the figure
        if isinstance(quotinent, int):
            dim_1 = 2
            dim_2 = quotinent
        else:
            dim_1 = np.ceil(quotinent)
            dim_2 = 3

        palette = sns.color_palette('deep',
np.unique(algorithm[estimators[i]['estimator'].__name__]).max() + 1)
```

```
            colors = [palette[x] if x >= 0 else (0.0, 0.0, 0.0) for x in
                      algorithm[estimators[i]['estimator'].__name__]]

            plt.subplot(dim_1, dim_2, i + 1)
            plt.scatter(X[:, 0], X[:, 1], c=colors, **plot_kwargs)
            plt.title('{}
Clusters'.format(str(estimators[i]['estimator'].__name__)), fontsize=8)

        plt.show()
```

查看用法：

```
plot_kwargs = {'s': 12, 'linewidths': 0.1}
unsupervised_learner.plot_all_clusters(estimators,
unsupervised_learner.predictions, X, plot_kwargs)
```

以上代码输出如图 4-15 所示。

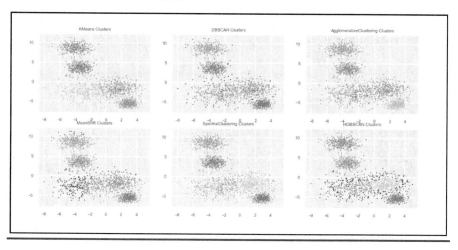

图 4-15　各种聚类

高维数据集视觉化
Visualizing High-dimensional Datasets

超过三个维度的数据集如何视觉化查看？为了数据集的视觉化效果，最多只能有三个维度。如果超出三个维度，则需要使用特定方法降维。一般使用**主成分分析法**（PCA）或 t-SNE 算法。

使用以下代码加载 Breast Cancer Wisconsin Diagnostics（威斯康星乳腺癌诊断）

数据集,这是机器学习教材中常用到的数据集:

```python
# Wisconsin Breast Cancer Diagnostic Dataset
from sklearn.datasets import load_breast_cancer
import pandas as pd

data = load_breast_cancer()
X = data.data

df = pd.DataFrame(data.data, columns=data.feature_names)
df.head()
```

控制台输出如下:

```
   mean radius  mean texture  mean perimeter  mean area  mean smoothness \
0        17.99         10.38          122.80     1001.0          0.11840
1        20.57         17.77          132.90     1326.0          0.08474
2        19.69         21.25          130.00     1203.0          0.10960
3        11.42         20.38           77.58      386.1          0.14250
4        20.29         14.34          135.10     1297.0          0.10030

...

   mean fractal dimension ... worst radius \
0                 0.07871 ...        25.38
1                 0.05667 ...        24.99
2                 0.05999 ...        23.57
3                 0.09744 ...        14.91
4                 0.05883 ...        22.54

...

   worst fractal dimension
0                  0.11890
1                  0.08902
2                  0.08758
3                  0.17300
4                  0.07678
```

了解一个病人的肿瘤,有 30 种不同的特征可用。

df.describe() 会显示出每个特征的描述统计数据,代码如下:

df.describe()

```
       mean radius  mean texture  mean perimeter   mean area \
count   569.000000    569.000000      569.000000  569.000000
mean     14.127292     19.289649       91.969033  654.889104
std       3.524049      4.301036       24.298981  351.914129
```

```
       min   6.981000   9.710000   43.790000   143.500000
       25%  11.700000  16.170000   75.170000   420.300000
       50%  13.370000  18.840000   86.240000   551.100000
       75%  15.780000  21.800000  104.100000   782.700000
       max  28.110000  39.280000  188.500000  2501.000000

       ...

              mean symmetry  mean fractal dimension ...  \
       count     569.000000              569.000000 ...
       mean        0.181162                0.062798 ...
       std         0.027414                0.007060 ...
       min         0.106000                0.049960 ...
       25%         0.161900                0.057700 ...
       50%         0.179200                0.061540 ...
       75%         0.195700                0.066120 ...
       max         0.304000                0.097440 ...

       ...

              worst concave points  worst symmetry  worst fractal dimension
       count            569.000000      569.000000               569.000000
       mean               0.114606        0.290076                 0.083946
       std                0.065732        0.061867                 0.018061
       min                0.000000        0.156500                 0.055040
       25%                0.064930        0.250400                 0.071460
       50%                0.099930        0.282200                 0.080040
       75%                0.161400        0.317900                 0.092080
       max                0.291000        0.663800                 0.207500
       [8 rows x 30 columns]
```

我们看看缩放前后的结果。

主成分分析实践
Principal Component Analysis in Action

以下代码展示如何应用有两个成分的 PCA，并给出可查看的结果：

```
# PCA
from sklearn.decomposition import PCA

pca = PCA(n_components=2, whiten=True)
pca = pca.fit_transform(df)

plt.scatter(pca[:, 0], pca[:, 1], c=data.target, cmap="RdBu_r",
edgecolor="Red", alpha=0.35)
plt.colorbar()
```

```
plt.title('PCA, n_components=2')
plt.show()
```

以上代码输出如图 4-16 所示。

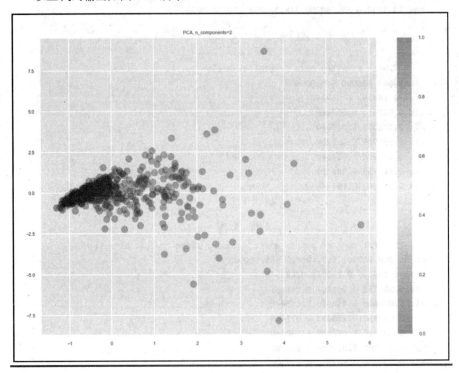

图 4-16　PCA 两成分分布图

可见，红色类（深灰）在某一个区域特别密集，很难分类。方差会导致视图变形，缩放可以改善，代码如下：

```
# Preprocess data.
scaler = StandardScaler()
scaler.fit(df)
preprocessed_data = scaler.transform(df)
scaled_features_df = pd.DataFrame(preprocessed_data, index=df.index,
columns=df.columns)
```

使用 StandardScaler 预处理数据之后，数据集有单元方差：

```
scaled_features_df.describe()

            mean radius  mean texture  mean perimeter  mean area  \
count       5.690000e+02  5.690000e+02  5.690000e+02  5.690000e+02
```

```
mean  -3.162867e-15 -6.530609e-15 -7.078891e-16 -8.799835e-16
std    1.000880e+00  1.000880e+00  1.000880e+00  1.000880e+00
min   -2.029648e+00 -2.229249e+00 -1.984504e+00 -1.454443e+00
25%   -6.893853e-01 -7.259631e-01 -6.919555e-01 -6.671955e-01
50%   -2.150816e-01 -1.046362e-01 -2.359800e-01 -2.951869e-01
75%    4.693926e-01  5.841756e-01  4.996769e-01  3.635073e-01
max    3.971288e+00  4.651889e+00  3.976130e+00  5.250529e+00

...

       mean symmetry mean fractal dimension ... \
count   5.690000e+02   5.690000e+02 ...
mean   -1.971670e-15  -1.453631e-15 ...
std     1.000880e+00   1.000880e+00 ...
min    -2.744117e+00  -1.819865e+00 ...
25%    -7.032397e-01  -7.226392e-01 ...
50%    -7.162650e-02  -1.782793e-01 ...
75%     5.307792e-01   4.709834e-01 ...
max     4.484751e+00   4.910919e+00 ...

...

       worst concave points worst symmetry worst fractal dimension
count         5.690000e+02    5.690000e+02          5.690000e+02
mean         -1.412656e-16   -2.289567e-15          2.575171e-15
std           1.000880e+00    1.000880e+00          1.000880e+00
min          -1.745063e+00   -2.160960e+00         -1.601839e+00
25%          -7.563999e-01   -6.418637e-01         -6.919118e-01
50%          -2.234689e-01   -1.274095e-01         -2.164441e-01
75%           7.125100e-01    4.501382e-01          4.507624e-01
max           2.685877e+00    6.046041e+00          6.846856e+00
[8 rows x 30 columns]
```

应用 PCA 看看前两个主成分是否能区分开标签：

```
# PCA
from sklearn.decomposition import PCA

pca = PCA(n_components=2, whiten=True)
pca = pca.fit_transform(scaled_features_df)

plt.scatter(pca[:, 0], pca[:, 1], c=data.target, cmap="RdBu_r",
edgecolor="Red", alpha=0.35)
plt.colorbar()
plt.title('PCA, n_components=2')
plt.show()
```

以上代码可得到图 4-17。

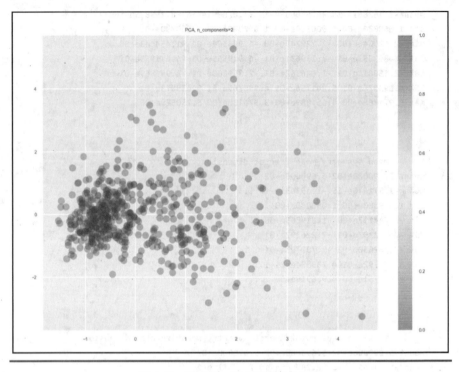

图 4-17　缩放后，PCA 两成分分布图

有意思，大多数不同标签的样本使用前两个主成分就可以区分了。

t-SNE 实践
t-SNE in Action

高维数据可视化也可尝试使用 t-SNE。首先，对原始数据应用 TSNE：

```
# TSNE
from sklearn.manifold import TSNE

tsne = TSNE(verbose=1, perplexity=40, n_iter=4000)
tsne = tsne.fit_transform(df)
```

控制台输出如下：

```
[t-SNE] Computing 121 nearest neighbors...
[t-SNE] Indexed 569 samples in 0.000s...
[t-SNE] Computed neighbors for 569 samples in 0.010s...
[t-SNE] Computed conditional probabilities for sample 569 / 569
[t-SNE] Mean sigma: 33.679703
```

```
[t-SNE] KL divergence after 250 iterations with early exaggeration:
48.886528
[t-SNE] Error after 1600 iterations: 0.210506
```

将结果绘制成图，代码如下：

```
plt.scatter(tsne[:, 0], tsne[:, 1], c=data.target, cmap="winter",
edgecolor="None", alpha=0.35)
plt.colorbar()
plt.title('t-SNE')
plt.show()
```

以上代码得到图4-18。

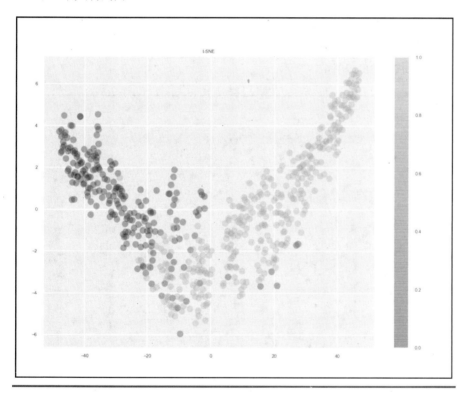

图4-18 t-SNE 图

在缩放后的数据中应用TSNE，代码如下：

```
tsne = TSNE(verbose=1, perplexity=40, n_iter=4000)
tsne = tsne.fit_transform(scaled_features_df)
```

控制台输出如下：

```
[t-SNE] Computing 121 nearest neighbors...
[t-SNE] Indexed 569 samples in 0.001s...
[t-SNE] Computed neighbors for 569 samples in 0.018s...
[t-SNE] Computed conditional probabilities for sample 569 / 569
[t-SNE] Mean sigma: 1.522404
[t-SNE] KL divergence after 250 iterations with early exaggeration:
66.959343
[t-SNE] Error after 1700 iterations: 0.875110
```

结果绘制如下:

```
plt.scatter(tsne[:, 0], tsne[:, 1], c=data.target, cmap="winter",
edgecolor="None", alpha=0.35)
plt.colorbar()
plt.title('t-SNE')
plt.show()
```

以上代码输出如图 4-19 所示。

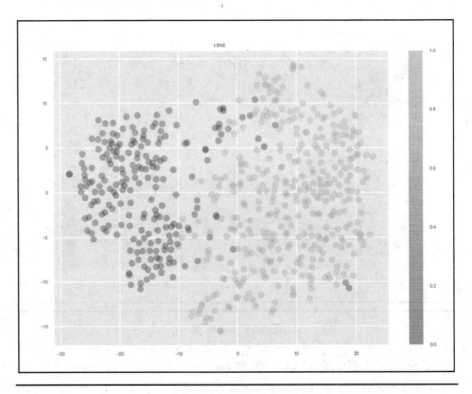

图 4-19　缩放后的 t-SNE 图

简单成分叠加以改善流水线
Adding Simple Components Together to Improve the Pipeline

调整 fit_predict 方法,在流水线中加入一个解析器,必要时可对高维数据进行可视化:

```python
def fit_predict(self, X, y=None, scaler=True, decomposer={'name': PCA,
'args':[], 'kwargs': {'n_components': 2}}):
    """
    fit_predict will train given estimator(s) and predict cluster
membership for each sample
    """

    shape = X.shape
    df_type = isinstance(X, pd.core.frame.DataFrame)

    if df_type:
        column_names = X.columns
        index = X.index

    if scaler == True:
        from sklearn.preprocessing import StandardScaler
        scaler = StandardScaler()
        X = scaler.fit_transform(X)

        if df_type:
            X = pd.DataFrame(X, index=index, columns=column_names)

    if decomposer is not None:
        X = decomposer['name'](*decomposer['args'],
**decomposer['kwargs']).fit_transform(X)

        if df_type:
            if decomposer['name'].__name__ == 'PCA':
                X = pd.DataFrame(X, index=index, columns=['component_' +
str(i + 1) for i in
range(decomposer['kwargs']['n_components'])])
            else:
                X = pd.DataFrame(X, index=index, columns=['component_1',
'component_2'])

        # if dimensionality reduction is applied, then n_components will be
set accordingly in hyperparameter configuration
        for estimator in self.estimators:
            if 'n_clusters' in estimator['kwargs'].keys():
                if decomposer['name'].__name__ == 'PCA':
                    estimator['kwargs']['n_clusters'] =
decomposer['kwargs']['n_components']
                else:
```

```python
                    estimator['kwargs']['n_clusters'] = 2

        # This dictionary will hold predictions for each estimator
        predictions = []
        performance_metrics = {}

        for estimator in self.estimators:
            labels = estimator['estimator'](*estimator['args'],
**estimator['kwargs']).fit_predict(X)
            estimator['estimator'].n_clusters_ = len(np.unique(labels))
            metrics =
self._get_cluster_metrics(estimator['estimator'].__name__,
estimator['estimator'].n_clusters_, X, labels, y)
            predictions.append({estimator['estimator'].__name__: labels})
            performance_metrics[estimator['estimator'].__name__] = metrics

        self.predictions = predictions
        self.performance_metrics = performance_metrics

        return predictions, performance_metrics
```

现在对数据集应用fit_predict方法。以下代码为用法示例：

```python
from sklearn.cluster import KMeans, DBSCAN, AgglomerativeClustering,
MeanShift, estimate_bandwidth, SpectralClustering
from hdbscan import HDBSCAN

from sklearn.datasets import load_breast_cancer

data = load_breast_cancer()
X = data.data
y = data.target

# Necessary for bandwidth
bandwidth = estimate_bandwidth(X, quantile=0.1, n_samples=100)

estimators = [{'estimator': KMeans, 'args': (), 'kwargs': {'n_clusters':
5}},
                        {'estimator': DBSCAN, 'args': (), 'kwargs':
{'eps': 0.3}},
                        {'estimator': AgglomerativeClustering, 'args': (),
'kwargs': {'n_clusters': 5, 'linkage': 'ward'}},
                        {'estimator': MeanShift, 'args': (), 'kwargs':
{'cluster_all': False, "bandwidth": bandwidth, "bin_seeding": True}},
                        {'estimator': SpectralClustering, 'args': (),
'kwargs': {'n_clusters':5}},
                        {'estimator': HDBSCAN, 'args': (), 'kwargs':
{'min_cluster_size':15}}]

unsupervised_learner = Unsupervised_AutoML(estimators)

predictions, performance_metrics = unsupervised_learner.fit_predict(X, y,
decomposer=None)
```

自动化无监督学习是一个高度试验性的过程，对数据了解有限的时候，尤为如此。作为练习，你可以扩展 Unsupervised_AutoML 类，每种算法可多尝试一些超参数配置，并将结果视觉化。

4.8　总结
Summary

本章从多个角度学习了如何为已知问题选择合适的机器学习流水线。

计算复杂度、训练时间和推理时间的区别、线性与非线性以及算法和必要特征转换等都是要考虑的因素，从这些角度评估数据是有益的。

各种用例练习之后，能更好地理解选择合适的模型和机器学习流水线的原理。本章仅接触了一些皮毛，但这也是一个不错的扩展技能的起点。

第 5 章会学习超参数优化并介绍更高阶的概念，如贝叶斯超参数优化。

第 5 章
超参数优化
Hyperparameter Optimization

auto-sklearn 库中使用**贝叶斯优化**（Bayesian optimization）进行**机器学习流水线**的超参数优化。本章会学习贝叶斯优化的内在原理，现在先重温一下数学优化的基础知识。

简言之，优化是选择把函数值最小化或最大化成最佳变量。如果目标是最小化，则该函数又称**损失函数**（loss function）或**代价函数**（cost function）。如果目标是最大化，则该函数又称**利用函数**（utility function）或**拟合函数**（fitness function）。比如，构建机器学习模型时，损失函数有助于在训练阶段将预测误差降到最小。

如果广义地看这整个过程，则会有许多关联变量。

第一，要系统地决定问题的类型，确定它是无监督、监督、半监督或是强化学习问题。根据数据大小和复杂度来确定硬件和软件的配置，然后选择适合实验使用的编程语言或库。从可用的**转换器**（transformer）和**估算器**（estimator）集合中选择一些在训练、验证和测试阶段会用到的子集。

这些都称为**配置参数**（configuration parameter），为机器学习流水线的开发设置场景。

第二，在训练阶段，要整理出转换器和估算器自身的参数，如线性模型中的因子，或者创建多项式和交互特征的级别参数。

比如，机器学习算法通常分为参数型的或非参数型的。若一个算法有固定数量的参数，则说明函数形状已知，是参数型的；否则，就是**非参数型的**（non-parametric），函数的形状由数据决定。

第三，除了参数，训练开始前还要设置超参数，以便引导转换器和**估算器**参数的估算。

超参数尤其重要，因为流水线的性能取决于超参数。由于超参数众多，且每个超参数都有一系列值可取，所以你会发现超参数需要优化。

已知超参数和取值范围，哪个搜索空间？如何高效地找出性能最好的机器学习流水线呢？在实践中，性能最好的机器学习流水线是交叉检验得分最高的那个。

本章涵盖如下主题。

- 机器学习实验的配置空间。
- 机器学习模型参数和超参数。
- 什么是热启动，以及怎样助力参数优化。
- 基于贝叶斯的超参数优化。
- 示例系统。

5.1 技术要求
Technical Requirements

你可以在本书的 GitHub 库的 Chapter 05 文件夹里找到全部代码示例。

5.2 超参数
Hyperparameters

为了更好地理解这个过程，我们先从简单的贝叶斯函数开始介绍，这个函数有 3 个全局极小值：

$$f(x) = a(x_2 - bx_1^2 + cx_1 - r)^2 + s(1-t)\cos(x_1) + s$$

以下代码展示了贝叶斯函数的最小化：

```python
import numpy as np

def branin(x):

    # Branin function has 2 dimensions and it has 3 global mimima
    x1 = x[0]
    x2 = x[1]
    # Global minimum is f(x*)=0.397887 at points (-pi, 12.275), (pi,2.275) and (9.42478, 2.475)

    # Recommended values of a, b, c, r, s and t for Branin function
    a = 1
    b = 5.1 / (4 * np.pi**2)
    c = 5. / np.pi
    r = 6.
    s = 10.
    t = 1 / (8 * np.pi)

    # Calculating separate parts of the function first for verbosity
    p1 = a * (x2 - (b * x1**2) + (c * x1) - r)**2
    p2 = s * (1-t) * np.cos(x1)
    p3 = s

    # Calculating result
    ret = p1 + p2 + p3

    return ret

# minimize function from scipy.optimize will minimize a scalar function
# with one or more variables
from scipy.optimize import minimize

x = [5.6, 3.2]

res = minimize(branin, x)

print(res)
```

执行以上代码，输出如下：

```
fun: 0.3978873577297417
hess_inv: array([[0.10409341, -0.0808961],
[-0.0808961, 0.56160622]])
jac: array([3.57627869e-07, -1.19209290e-07])
message: 'Optimization terminated successfully.'
nfev: 36
nit: 5
njev: 9
status: 0
success: True
x: array([3.14159268, 2.27499994])
```

优化成功结束，在贝叶斯函数的输出（3.14159268, 2.27499994）中可找到全局最小值。这个优化问题还有许多种解决方案，如 BFGS、L-BFGS-B 和 SLSQP，特性各不相同，如一致性和复杂性不同。通过示例练习，可熟悉其中的一些解决方案，为未来的探索开辟空间。

我们回顾一下机器学习问题优化的基础知识。大多数机器学习最终简化为以下公式：

$$L = \sum_{i=1}^{N}(y_i - f(x_i|w,b))^2 + \alpha\sum_{j=1}^{D}w_j^2 + \beta\sum_{j=1}^{D}|w_j|$$

该公式中有损失函数和正则化条件，以防止过度拟合。权重，用 w 表示，是训练过程中尝试要学习的参数，是前面提到的学习算法的参数。除此之外，通常还要定义超参数，如学习速率和提前停止条件，这些都会影响学习行为。

你是否注意到损失函数中的 α 和 β？这是训练前要设置的参数，也是超参数。

超参数有助于在模型偏差和模型方差之间保持健康平衡。

我们看一个在 sklearn 中关于估算器参数的简单例子：

```
from sklearn.linear_model import LogisticRegression

log_reg = LogisticRegression()
log_reg.get_params()
```

输出如下：

```
{'C': 1.0,
 'class_weight': None,
 'dual': False,
 'fit_intercept': True,
 'intercept_scaling': 1,
 'max_iter': 100,
```

```
'multi_class': 'ovr',
'n_jobs': 1,
'penalty': 'l2',
'random_state': None,
'solver': 'liblinear',
'tol': 0.0001,
'verbose': 0,
'warm_start': False}
```

这里有 14 个超参数，想想各种可能的组合，你就知道搜索空间有多大了。我们现在的目标是从全部的超参数集中找出交叉检验得分最高的超参数。

LogisticRegression 的一个重要的超参数是 C，它用于控制正则化的强度。此值与正则化的强度负相关，即 C 值越大，正则化越弱。

即使你是研究算法方面的专家，也要不断试验才能设置合适的超参数，而且也受限于从业者的经验。我们要找出比启发式更好的办法，找出一套接近最优或最优的超参数。

比如，可以使用 GridSearchCV 或 RandomizedSearchCV 找出 sklearn 中的超参数空间。

- GridSearchCV 从已知的超参数和取值范围中生成候选集。假设有如下参数网格：

```
# Hyperparameters
param_grid = [ {'C': [0.001, 0.01, 0.1, 1, 10, 20, 50, 100],
                'penalty': ['l1', 'l2']} ]
```

那么，GridSearchCV 会生成如下参数：

```
'params': [{'C': 0.001, 'penalty': 'l1'},
 {'C': 0.001, 'penalty': 'l2'},
 {'C': 0.01, 'penalty': 'l1'},
 {'C': 0.01, 'penalty': 'l2'},
 {'C': 0.1, 'penalty': 'l1'},
 {'C': 0.1, 'penalty': 'l2'},
 {'C': 1, 'penalty': 'l1'},
 {'C': 1, 'penalty': 'l2'},
 {'C': 10, 'penalty': 'l1'},
 {'C': 10, 'penalty': 'l2'},
 {'C': 20, 'penalty': 'l1'},
 {'C': 20, 'penalty': 'l2'},
 {'C': 50, 'penalty': 'l1'},
 {'C': 50, 'penalty': 'l2'},
 {'C': 100, 'penalty': 'l1'},
 {'C': 100, 'penalty': 'l2'}]
```

GridSearchCV 会穷举搜索，找出交叉检验最高分。

- RandomizedSearchCV 搜索的方式与 GridSearchCV 的不同。它不是穷举搜索超参数空间，而是从指定的分布中进行参数设置采样。采用如下方式构建参数网格：

```
# Hyperparameters
param_grid = {'C': sp_randint(1, 100),
              'penalty': ['l1', 'l2']}
```

你是否注意到 sp_randint？它可以使 RandomizedSearchCV 从均匀分布中随机抽取变量，创建的参数如下：

```
'params': [{'C': 6, 'penalty': 'l2'},
  {'C': 97, 'penalty': 'l2'},
  {'C': 92, 'penalty': 'l2'},
  {'C': 62, 'penalty': 'l1'},
  {'C': 63, 'penalty': 'l2'},
  {'C': 5, 'penalty': 'l2'},
  {'C': 7, 'penalty': 'l1'},
  {'C': 45, 'penalty': 'l1'},
  {'C': 77, 'penalty': 'l2'},
  {'C': 12, 'penalty': 'l1'},
  {'C': 72, 'penalty': 'l2'},
  {'C': 28, 'penalty': 'l1'},
  {'C': 7, 'penalty': 'l2'},
  {'C': 65, 'penalty': 'l1'},
  {'C': 32, 'penalty': 'l1'},
  {'C': 84, 'penalty': 'l1'},
  {'C': 27, 'penalty': 'l1'},
  {'C': 12, 'penalty': 'l1'},
  {'C': 21, 'penalty': 'l1'},
  {'C': 65, 'penalty': 'l1'}],
```

分别查看 GridSearchCV 和 RandomizedSearchCV 的用法示例。

以下代码片段展示的是 GridSearchCV：

```
from sklearn.linear_model import LogisticRegression

log_reg = LogisticRegression()

# Hyperparameters
param_grid = {'C': [0.001, 0.01, 0.1, 1, 10, 20, 50, 100],
              'penalty': ['l1', 'l2']}

from sklearn.model_selection import GridSearchCV

n_folds = 5
estimator = GridSearchCV(log_reg,param_grid, cv=n_folds)
```

```python
from sklearn import datasets
iris = datasets.load_iris()
X = iris.data
Y = iris.target

estimator.fit(X, Y)
```

输出如下：

```
GridSearchCV(cv=5, error_score='raise',
       estimator=LogisticRegression(C=1.0, class_weight=None, dual=False,
fit_intercept=True,
          intercept_scaling=1, max_iter=100, multi_class='ovr', n_jobs=1,
          penalty='l2', random_state=None, solver='liblinear', tol=0.0001,
          verbose=0, warm_start=False),
       fit_params=None, iid=True, n_jobs=1,
       param_grid=[{'C': [0.001, 0.01, 0.1, 1, 10, 20, 50, 100], 'penalty': ['l1', 'l2']}],
       pre_dispatch='2*n_jobs', refit=True, return_train_score=True,
       scoring=None, verbose=0)
```

训练结束后，可查看性能最好的估算器配置：

estimator.best_estimator_

以上代码输出如下：

```
LogisticRegression(C=10, class_weight=None, dual=False, fit_intercept=True,
intercept_scaling=1, max_iter=100, multi_class='ovr', n_jobs=1,
penalty='l1', random_state=None, solver='liblinear', tol=0.0001,
verbose=0, warm_start=False)
```

也可查看最高分：

estimator.best_score_

输出如下：

0.98

查看 cv_results_ 可看到全部结果：

estimator.cv_results_

每次训练的各种指标都会显示如下：

```
{'mean_fit_time': array([0.00039144, 0.00042701, 0.00036378, 0.00043044,
0.00145531,
        0.00046387, 0.00670047, 0.00056334, 0.00890565, 0.00064907,
        0.00916181, 0.00063758, 0.01110044, 0.00076027, 0.01196856,
        0.00084472]),
 'mean_score_time': array([0.00017729, 0.00018134, 0.00016704, 0.00016623,
0.00017071,
```

```
                0.00016556, 0.00024438, 0.00017123, 0.00020232, 0.00018559,
                0.00020504, 0.00016532, 0.00024428, 0.00019045, 0.00023465,
                0.00023274]),
 'mean_test_score': array([0.33333333, 0.40666667, 0.33333333, 0.66666667,
0.77333333,
                0.82      , 0.96      , 0.96      , 0.98      , 0.96666667,
                0.96666667, 0.96666667, 0.96666667, 0.97333333, 0.96      ,
                0.98      ]),
 'mean_train_score': array([0.33333333, 0.40166667, 0.33333333, 0.66666667,
0.775      ,
                0.83166667, 0.96333333, 0.96333333, 0.97333333, 0.97333333,
                0.97333333, 0.97666667, 0.975     , 0.97833333, 0.975     ,
                0.98      ]),
 'param_C': masked_array(data=[0.001, 0.001, 0.01, 0.01, 0.1, 0.1, 1, 1,
10, 10, 20,
                   20, 50, 50, 100, 100],
             mask=[False, False, False, False, False, False, False, False,
                   False, False, False, False, False, False, False,
False],
        fill_value='?',
             dtype=object),
 'param_penalty': masked_array(data=['l1', 'l2', 'l1', 'l2', 'l1', 'l2',
'l1', 'l2', 'l1',
                   'l2', 'l1', 'l2', 'l1', 'l2', 'l1', 'l2'],
             mask=[False, False, False, False, False, False, False, False,
                   False, False, False, False, False, False, False,
False],
        fill_value='?',
             dtype=object),
 'params': [{'C': 0.001, 'penalty': 'l1'},
  {'C': 0.001, 'penalty': 'l2'},
  {'C': 0.01, 'penalty': 'l1'},
  {'C': 0.01, 'penalty': 'l2'},
  {'C': 0.1, 'penalty': 'l1'},
  {'C': 0.1, 'penalty': 'l2'},
  {'C': 1, 'penalty': 'l1'},
  {'C': 1, 'penalty': 'l2'},
  {'C': 10, 'penalty': 'l1'},
  {'C': 10, 'penalty': 'l2'},
  {'C': 20, 'penalty': 'l1'},
  {'C': 20, 'penalty': 'l2'},
  {'C': 50, 'penalty': 'l1'},
  {'C': 50, 'penalty': 'l2'},
  {'C': 100, 'penalty': 'l1'},
  {'C': 100, 'penalty': 'l2'}],
 'rank_test_score': array([15, 14, 15, 13, 12, 11, 8, 8, 1, 4, 4, 4, 4, 3,
8, 1],
        dtype=int32),
 'split0_test_score': array([0.33333333, 0.36666667, 0.33333333,
0.66666667, 0.7       ,
                0.76666667, 1.        , 1.        , 1.        , 1.        ,
                1.        , 1.        , 1.        , 1.        , 0.96666667,
                1.        ]),
```

```
 'split0_train_score': array([0.33333333, 0.41666667, 0.33333333,
0.66666667, 0.775 ,
        0.825 , 0.95 , 0.95 , 0.95 , 0.96666667,
        0.95 , 0.975 , 0.95833333, 0.975 , 0.95833333,
        0.975 ]),
 'split1_test_score': array([0.33333333, 0.46666667, 0.33333333,
0.66666667, 0.8 ,
        0.86666667, 0.96666667, 0.96666667, 1. , 1. ,
        0.96666667, 1. , 0.96666667, 1. , 0.96666667,
        1. ]),
 'split1_train_score': array([0.33333333, 0.35833333, 0.33333333,
0.66666667, 0.775 ,
        0.825 , 0.95833333, 0.96666667, 0.975 , 0.96666667,
        0.975 , 0.975 , 0.975 , 0.975 , 0.975 ,
        0.975 ]),
 'split2_test_score': array([0.33333333, 0.36666667, 0.33333333,
0.66666667, 0.8 ,
        0.83333333, 0.93333333, 0.93333333, 0.96666667, 0.93333333,
        0.93333333, 0.93333333, 0.93333333, 0.93333333, 0.93333333,
        0.96666667]),
 'split2_train_score': array([0.33333333, 0.41666667, 0.33333333,
0.66666667, 0.76666667,
        0.83333333, 0.96666667, 0.96666667, 0.975 , 0.975 ,
        0.975 , 0.98333333, 0.975 , 0.98333333, 0.975 ,
        0.98333333]),
 'split3_test_score': array([0.33333333, 0.46666667, 0.33333333,
0.66666667, 0.8 ,
        0.83333333, 0.9 , 0.9 , 0.93333333, 0.9 ,
        0.93333333, 0.9 , 0.93333333, 0.93333333, 0.93333333,
        0.93333333]),
 'split3_train_score': array([0.33333333, 0.39166667, 0.33333333,
0.66666667, 0.775 ,
        0.84166667, 0.975 , 0.975 , 0.99166667, 0.98333333,
        0.99166667, 0.98333333, 0.99166667, 0.98333333, 0.99166667,
        0.99166667]),
 'split4_test_score': array([0.33333333, 0.36666667, 0.33333333,
0.66666667, 0.76666667,
        0.8 , 1. , 1. , 1. , 1. ,
        1. , 1. , 1. , 1. , 1. ,
        1. ]),
 'split4_train_score': array([0.33333333, 0.425 , 0.33333333, 0.66666667,
0.78333333,
        0.83333333, 0.96666667, 0.95833333, 0.975 , 0.975 ,
        0.975 , 0.96666667, 0.975 , 0.975 , 0.975 ,
        0.975 ]),
 'std_fit_time': array([7.66660734e-05, 3.32198455e-05, 1.98168153e-05,
6.91923414e-06,
        4.74922317e-04, 2.65661212e-05, 1.03221712e-03, 3.79795334e-05,
        1.86899641e-03, 8.53752397e-05, 1.93386463e-03, 2.95752073e-05,
        2.91377734e-03, 5.70420424e-05, 3.59721435e-03, 9.67829087e-05]),
 'std_score_time': array([1.28883712e-05, 2.39771817e-05, 4.81959487e-06,
2.47955322e-06,
        1.34236224e-05, 2.41545203e-06, 5.64869920e-05, 8.94803700e-06,
```

```
                    4.10209125e-05, 3.35513820e-05, 3.04168290e-05, 2.87924369e-06,
                    4.91685012e-05, 1.62987656e-05, 4.23611246e-05, 7.26868455e-05]),
 'std_test_score': array([0.    , 0.04898979, 0.    , 0.    , 0.03887301,
                    0.03399346, 0.03887301, 0.03887301, 0.02666667, 0.0421637 ,
                    0.02981424, 0.0421637 , 0.02981424, 0.03265986, 0.02494438,
                    0.02666667]),
 'std_train_score': array([0.    , 0.02438123, 0.    , 0.    , 0.00527046,
                    0.0062361 , 0.00849837, 0.00849837, 0.01333333, 0.0062361 ,
                    0.01333333, 0.0062361 , 0.01054093, 0.00408248, 0.01054093,
                    0.00666667])}
```

下面查看 RandomizedSearchCV 是如何使用的:

```
from sklearn.model_selection import RandomizedSearchCV
from scipy.stats import randint as sp_randint

# Hyperparameters
param_grid = {'C': sp_randint(1, 100),
              'penalty': ['l1', 'l2']}

n_iter_search = 20
n_folds = 5
estimator = RandomizedSearchCV(log_reg, param_distributions=param_grid,
n_iter=n_iter_search, cv=n_folds)

estimator.fit(X, Y)
```

以上代码输出结果与 GridSearchCV 的相似:

```
RandomizedSearchCV(cv=5, error_score='raise',
        estimator=LogisticRegression(C=1.0, class_weight=None,
dual=False, fit_intercept=True,
          intercept_scaling=1, max_iter=100, multi_class='ovr', n_jobs=1,
          penalty='l2', random_state=None, solver='liblinear', tol=0.0001,
          verbose=0, warm_start=False),
          fit_params=None, iid=True, n_iter=20, n_jobs=1,
          param_distributions={'C':
<scipy.stats._distn_infrastructure.rv_frozen object at 0x1176d4c88>,
'penalty': ['l1', 'l2']},
          pre_dispatch='2*n_jobs', random_state=None, refit=True,
          return_train_score=True, scoring=None, verbose=0)
```

同样查看 best_estimator_:

```
estimator.best_estimator_
```

以上代码输出如下:

```
LogisticRegression(C = 95, class_weight=None, dual=False,
fit_intercept=True,
          intercept_scaling=1, max_iter=100, multi_class='ovr', n_jobs=1,
          penalty ='l2', random_state=None, solver='liblinear', tol=0.0001,
          verbose = 0, warm_start=False)
```

estimator.best_score_ 显示如下：

0.98

RandomizedSearchCV 的最高分也一样。要注意的是，最佳性能估算器的设置有 C=95，这个值很难出现，因为人们在人工构建参数网格时一般会尝试整数，如 10、100 或 1000 等。

类似地，也可使用 estimator.cv_results_ 查看交叉检验结果：

```
{'mean_fit_time': array([0.0091342 , 0.00065241, 0.00873041, 0.00068126,
0.00082703,
        0.01093817, 0.00067267, 0.00961967, 0.00883713, 0.00069351,
        0.01048965, 0.00068388, 0.01074204, 0.0090354 , 0.00983639,
        0.01081419, 0.01014266, 0.00067706, 0.01015086, 0.00067825]),
 'mean_score_time': array([0.00026116, 0.0001647 , 0.00020576, 0.00017738,
0.00022368,
        0.00023923, 0.00016236, 0.00017295, 0.00026078, 0.00021319,
        0.00028219, 0.00018024, 0.00027289, 0.00025878, 0.00020723,
        0.00020337, 0.00023756, 0.00017438, 0.00028505, 0.0001936 ]),
 'mean_test_score': array([0.96666667, 0.97333333, 0.97333333, 0.98      ,
0.97333333,
        0.96666667, 0.97333333, 0.96666667, 0.98      , 0.97333333,
        0.96666667, 0.98      , 0.96666667, 0.96666667, 0.96666667,
        0.96666667, 0.96666667, 0.98      , 0.96666667, 0.96666667]),
 'mean_train_score': array([0.97333333, 0.97833333, 0.97333333, 0.98      ,
0.97833333,
        0.975     , 0.97833333, 0.975     , 0.97333333, 0.97833333,
        0.975     , 0.98      , 0.975     , 0.97333333, 0.975     ,
        0.975     , 0.975     , 0.97833333, 0.975     , 0.97666667]),
 'param_C': masked_array(data=[20, 53, 5, 95, 50, 71, 41, 43, 8, 30, 70,
91, 53, 15,
                    35, 41, 56, 82, 90, 27],
              mask=[False, False, False, False, False, False, False,
                    False, False, False, False, False, False, False,
                    False, False, False, False],
        fill_value='?',
            dtype=object),
 'param_penalty': masked_array(data=['l1', 'l2', 'l1', 'l2', 'l2', 'l1',
'l2', 'l1', 'l1',
                    'l2', 'l1', 'l2', 'l1', 'l1', 'l1', 'l1', 'l1', 'l2',
                    'l1', 'l2'],
              mask=[False, False, False, False, False, False, False, False,
                    False, False, False, False, False, False, False, False,
                    False, False, False, False],
        fill_value='?',
            dtype=object),
 'params': [{'C': 20, 'penalty': 'l1'},
  {'C': 53, 'penalty': 'l2'},
  {'C': 5, 'penalty': 'l1'},
  {'C': 95, 'penalty': 'l2'},
```

```
                       {'C': 50, 'penalty': 'l2'},
                       {'C': 71, 'penalty': 'l1'},
                       {'C': 41, 'penalty': 'l2'},
                       {'C': 43, 'penalty': 'l1'},
                       {'C': 8, 'penalty': 'l1'},
                       {'C': 30, 'penalty': 'l2'},
                       {'C': 70, 'penalty': 'l1'},
                       {'C': 91, 'penalty': 'l2'},
                       {'C': 53, 'penalty': 'l1'},
                       {'C': 15, 'penalty': 'l1'},
                       {'C': 35, 'penalty': 'l1'},
                       {'C': 41, 'penalty': 'l1'},
                       {'C': 56, 'penalty': 'l1'},
                       {'C': 82, 'penalty': 'l2'},
                       {'C': 90, 'penalty': 'l1'},
                       {'C': 27, 'penalty': 'l2'}],
 'rank_test_score': array([10, 5, 5, 1, 5, 10, 5, 10, 1, 5, 10, 1, 10, 10, 10, 10, 10,
        1, 10, 10], dtype=int32),
 'split0_test_score': array([1., 1., 1., 1., 1., 1., 1., 1., 1., 1., 1.,
        1., 1., 1., 1., 1.,
        1., 1., 1.]),
 'split0_train_score': array([0.95 , 0.975 , 0.95833333, 0.975 , 0.975,
        0.95833333, 0.975 , 0.95833333, 0.95833333, 0.975 ,
        0.95833333, 0.975 , 0.95833333, 0.95 , 0.95833333,
        0.95833333, 0.95833333, 0.975 , 0.95833333, 0.975 ]),
 'split1_test_score': array([0.96666667, 1. , 1. , 1. , 1. ,
        0.96666667, 1. , 0.96666667, 1. , 1. ,
        0.96666667, 1. , 0.96666667, 0.96666667, 0.96666667,
        0.96666667, 0.96666667, 1. , 0.96666667, 1. ]),
 'split1_train_score': array([0.975, 0.975, 0.975, 0.975, 0.975, 0.975,
        0.975, 0.975, 0.975,
        0.975, 0.975, 0.975, 0.975, 0.975, 0.975, 0.975, 0.975,
        0.975, 0.975]),
 'split2_test_score': array([0.93333333, 0.93333333, 0.93333333,
        0.96666667, 0.93333333,
        0.93333333, 0.93333333, 0.93333333, 0.96666667, 0.93333333,
        0.93333333, 0.96666667, 0.93333333, 0.93333333, 0.93333333,
        0.93333333, 0.93333333, 0.96666667, 0.93333333, 0.93333333]),
 'split2_train_score': array([0.975 , 0.98333333, 0.975 , 0.98333333,
        0.98333333,
        0.975 , 0.98333333, 0.975 , 0.975 , 0.98333333,
        0.975 , 0.98333333, 0.975 , 0.975 , 0.975 ,
        0.975 , 0.975 , 0.98333333, 0.975 , 0.98333333]),
 'split3_test_score': array([0.93333333, 0.93333333, 0.93333333,
        0.93333333, 0.93333333,
        0.93333333, 0.93333333, 0.93333333, 0.93333333, 0.93333333,
        0.93333333, 0.93333333, 0.93333333, 0.93333333, 0.93333333,
        0.93333333, 0.93333333, 0.93333333, 0.93333333, 0.9 ]),
 'split3_train_score': array([0.99166667, 0.98333333, 0.98333333,
        0.99166667, 0.98333333,
        0.99166667, 0.98333333, 0.99166667, 0.98333333, 0.98333333,
        0.99166667, 0.99166667, 0.99166667, 0.99166667, 0.99166667,
```

```
             0.99166667, 0.99166667, 0.98333333, 0.99166667, 0.98333333]),
 'split4_test_score': array([1., 1., 1., 1., 1., 1., 1., 1., 1., 1., 1.,
        1., 1., 1., 1., 1., 1.,
        1., 1., 1.]),
 'split4_train_score': array([0.975     , 0.975     , 0.975     , 0.975     , 0.975     ,
        0.975     , 0.975     , 0.975     , 0.975     , 0.975     ,
        0.975     , 0.975     , 0.975     , 0.975     , 0.975     ,
        0.975     , 0.975     , 0.975     , 0.975     , 0.96666667]),
 'std_fit_time': array([2.16497645e-03, 5.39653699e-05, 1.00355397e-03, 4.75298306e-05,
        9.75692490e-05, 2.63689357e-03, 7.04799517e-05, 2.52499464e-03,
        1.92020413e-03, 6.05031761e-05, 1.78589024e-03, 5.85074724e-05,
        2.28621528e-03, 2.19771432e-03, 1.96957384e-03, 3.06769107e-03,
        1.15194163e-03, 2.10475943e-05, 1.33958298e-03, 4.09795418e-05]),
 'std_score_time': array([4.62378644e-05, 1.66142000e-06, 3.40806829e-05, 1.73623737e-05,
        5.26490415e-05, 4.75790783e-05, 1.48510089e-06, 7.53432889e-06,
        3.86445261e-05, 8.16042958e-05, 4.98746594e-05, 1.93474877e-05,
        2.82650630e-05, 2.54787261e-05, 2.55031663e-05, 3.09080976e-05,
        2.99830109e-05, 7.89824294e-06, 2.02431836e-05, 4.25877252e-05]),
 'std_test_score': array([0.02981424, 0.03265986, 0.03265986, 0.02666667, 0.03265986,
        0.02981424, 0.03265986, 0.02981424, 0.02666667, 0.03265986,
        0.02981424, 0.02666667, 0.02981424, 0.02981424, 0.02981424,
        0.02981424, 0.02981424, 0.02666667, 0.02981424, 0.0421637 ]),
 'std_train_score': array([0.01333333, 0.00408248, 0.00816497, 0.00666667, 0.00408248,
        0.01054093, 0.00408248, 0.01054093, 0.00816497, 0.00408248,
        0.01054093, 0.00666667, 0.01054093, 0.01333333, 0.01054093,
        0.01054093, 0.01054093, 0.00408248, 0.01054093, 0.0062361 ])}
```

交叉检验结果看起来有点乱，但可将这些结果导入 pandas DataFrame：

```
import pandas as pd

df = pd.DataFrame(estimator.cv_results_)

df.head()
```

可看到如下记录：

```
  mean_fit_time mean_score_time mean_test_score mean_train_score param_C \
0      0.009134        0.000261        0.966667         0.973333      20
1      0.000652        0.000165        0.973333         0.978333      53
2      0.008730        0.000206        0.973333         0.973333       5
3      0.000681        0.000177        0.980000         0.980000      95
4      0.000827        0.000224        0.973333         0.978333      50
  param_penalty                    params  rank_test_score \
0            l1   {'C': 20, 'penalty': 'l1'}              10
1            l2   {'C': 53, 'penalty': 'l2'}               5
2            l1   {'C': 5, 'penalty': 'l1'}                5
3            l2   {'C': 95, 'penalty': 'l2'}               1
4            l2   {'C': 50, 'penalty': 'l2'}               5
  split0_test_score split0_train_score  ...  split2_test_score \
```

```
  0  1.0  0.950000 ...  0.933333
  1  1.0  0.975000 ...  0.933333
  2  1.0  0.958333 ...  0.933333
  3  1.0  0.975000 ...  0.966667
  4  1.0  0.975000 ...  0.933333
     split2_train_score split3_test_score split3_train_score  \
  0       0.975000          0.933333          0.991667
  1       0.983333          0.933333          0.983333
  2       0.975000          0.933333          0.983333
  3       0.983333          0.933333          0.991667
  4       0.983333          0.933333          0.983333
     split4_test_score split4_train_score std_fit_time std_score_time  \
  0       1.0            0.975           0.002165       0.000046
  1       1.0            0.975           0.000054       0.000002
  2       1.0            0.975           0.001004       0.000034
  3       1.0            0.975           0.000048       0.000017
  4       1.0            0.975           0.000098       0.000053
     std_test_score std_train_score
  0    0.029814       0.013333
  1    0.032660       0.004082
  2    0.032660       0.008165
  3    0.026667       0.006667
  4    0.032660       0.004082
  [5 rows x 22 columns]
```

你可以过滤 DataFrame，查看 mean_test_score 在哪里取得最大值：

```
df[df['mean_test_score'] == df['mean_test_score'].max()]
```

输出如下：

```
      mean_fit_time mean_score_time mean_test_score mean_train_score param_C  \
  3     0.000681      0.000177           0.98          0.980000         95
  8     0.008837      0.000261           0.98          0.973333          8
  11    0.000684      0.000180           0.98          0.980000         91
  17    0.000677      0.000174           0.98          0.978333         82
     param_penalty       params       rank_test_score  \
  3       l2     {'C': 95, 'penalty': 'l2'}     1
  8       l1     {'C': 8, 'penalty': 'l1'}      1
  11      l2     {'C': 91, 'penalty': 'l2'}     1
  17      l2     {'C': 82, 'penalty': 'l2'}     1
     split0_test_score split0_train_score ... split2_test_score  \
  3       1.0            0.975000       ...    0.966667
  8       1.0            0.958333       ...    0.966667
  11      1.0            0.975000       ...    0.966667
  17      1.0            0.975000       ...    0.966667
     split2_train_score split3_test_score split3_train_score  \
  3       0.983333          0.933333          0.991667
  8       0.975000          0.933333          0.983333
  11      0.983333          0.933333          0.991667
  17      0.983333          0.933333          0.983333
     split4_test_score split4_train_score std_fit_time std_score_time  \
  3       1.0            0.975           0.000048       0.000017
```

```
 8 1.0 0.975 0.001920 0.000039
11 1.0 0.975 0.000059 0.000019
17 1.0 0.975 0.000021 0.000008
    std_test_score std_train_score
 3 0.026667 0.006667
 8 0.026667 0.008165
11 0.026667 0.006667
17 0.026667 0.004082
[4 rows x 22 columns]
```

作为练习，你可以使用以下超参数为 `GradientBoostingClassifier` 创建一个参数网格，分别使用 `GridSearchCV` 和 `RandomizedSearchCV` 进行实验。

- `learning_rate` (default=0.1)：提升学习速率。
- `n_estimators` (default=100)：待拟合提升树的数量。
- `max_depth` (default=3)：最大树深。

5.3 热启动
Warm Start

在 AutoML 流水线中，超参数搜索空间有时增长非常迅速，在有限的时间和资源条件下，穷举搜索不切实际。完成这个任务就要使用更聪明的办法，尤其是要在使用复杂模型处理大数据集的时候。遇到这种情况，`GridSearchCV` 的实例穷举搜索是不可行的，`RandomizedSearchCV` 的参数随机抽取也无法在有限的时间里得出理想结果。

热启动的基本理念是利用从上一轮训练中得到的信息为下一轮训练找到更好的起点。

比如，`LogisticRegression` 有一个 `warm_start` 参数，默认设置是 `False`。以下示例展示了第一次的训练时间，设置为 `False` 时参数更新：

```
from sklearn.linear_model import LogisticRegression
log_reg = LogisticRegression(C=10, tol=0.00001)

from sklearn import datasets
iris = datasets.load_iris()
X = iris.data
Y = iris.target
```

```
from time import time
start = time()
log_reg.fit(X, Y)
end = time()
print("Time: {}".format(end - start))
# Time: 0.0009272098541259766

log_reg.set_params(C=20)
# LogisticRegression(C=100, class_weight=None, dual=False,
fit_intercept=True,
# intercept_scaling=1, max_iter=100, multi_class='ovr', n_jobs=1,
# penalty='l2', random_state=None, solver='liblinear', tol=0.0001,
# verbose=0, warm_start=False)

start = time()
log_reg.fit(X, Y)
end = time()
print("Time: {}".format(end - start))
# Time: 0.0012941360473632812
```

默认的 `LogisticRegression` 解决方案是 `liblinear`，每次新拟合的时候重新进行权重初始化，但其他解决方案，如 `lgfgs`、`newton-cg`、`sag` 和 `saga` 可利用 `warm_start` 使用之前拟合的信息减少计算时间。

以下代码演示了实践中如何使用热启动：

```
log_reg = LogisticRegression(C=10, solver='sag', warm_start=True,
max_iter=10000)

start = time()
log_reg.fit(X, Y)
end = time()
print("Time: {}".format(end - start))
# Time: 0.043714046478271484

log_reg.set_params(C=20)

start = time()
log_reg.fit(X, Y)
end = time()
print("Time: {}".format(end - start))
# Time: 0.020781755447387695
```

5.4 贝叶斯超参数优化
Bayesian-based Hyperparameter Tuning

模型超参数优化会用到几种方法，这些方法都统一来自于**序贯模型全局优化**

（sequential model-basedglobal optimization，SMBO）。

每当想到 GridSearchCV 或 RandomizedSearchCV 时，顺其自然地会感觉到这种交叉验证超参数的方法不是非常智能。两套预定义的超参数都在训练时验证，没办法从训练时获得的信息中受益。如果有办法从之前的模型性能超参数验证的迭代中学习，那可能更容易在下一次迭代中找出更高性能的超参数。

这是 SMBO 方法的初衷，贝叶斯超参数优化就是此类方法之一。

序贯模型算法配置（sequential model-based algorithm configuration，SMAC）是一个很好的库，用贝叶斯优化为已知机器学习算法配置超参数，使用便捷。

以下代码展示了如何使用 SMAC 优化开始使用的 branin 函数：

```
from smac.facade.func_facade import fmin_smac

x, cost, _ = fmin_smac(func=branin, # function
                       x0=[3.2, 4.5], # default configuration
                       bounds=[(-5, 10), (0, 15)], # limits
                       maxfun=500, # maximum number of evaluations
                       rng=3) # random seed


print(x, cost)
# [3.12848204 2.33810374] 0.4015064637498025
```

5.5 示例系统
An Example System

本节中，我们练习编写一个包装器函数来优化 XGBoost 算法超参数，提升它在 Breast Cancer Wisconsin 数据集中的性能：

```
# Importing necessary libraries
import numpy as np
from xgboost import XGBClassifier
from sklearn import datasets
from sklearn.model_selection import cross_val_score

# Importing ConfigSpace and different types of parameters
from smac.configspace import ConfigurationSpace
from ConfigSpace.hyperparameters import CategoricalHyperparameter, \
UniformFloatHyperparameter, UniformIntegerHyperparameter
```

```python
from ConfigSpace.conditions import InCondition

# Import SMAC-utilities
from smac.tae.execute_func import ExecuteTAFuncDict
from smac.scenario.scenario import Scenario
from smac.facade.smac_facade import SMAC

# Creating configuration space.
# Configuration space will hold all of your hyperparameters
cs = ConfigurationSpace()

# Defining hyperparameters and range of values that they can take
learning_rate = UniformFloatHyperparameter("learning_rate", 0.001, 0.1, default_value=0.1)
n_estimators = UniformIntegerHyperparameter("n_estimators", 100, 200, default_value=100)

# Adding hyperparameters to configuration space
cs.add_hyperparameters([learning_rate, n_estimators])
# Loading data set
wbc_dataset = datasets.load_breast_cancer()

# Creating function to cross validate XGBoost classifier given the configuration space
def xgboost_from_cfg(cfg):
    """ Creates a XGBoost based on a configuration and evaluates it on the
    Wisconsin Breast Cancer-dataset using cross-validation.

    Parameters:
    -----------
    cfg: Configuration (ConfigSpace.ConfigurationSpace.Configuration)
        Configuration containing the parameters.
        Configurations are indexable!
    Returns:
    --------
    A crossvalidated mean score for the svm on the loaded data-set.
    """

    cfg = {k: cfg[k] for k in cfg if cfg[k]}

    clf = XGBClassifier(**cfg, eval_metric='auc', early_stopping_rounds=50, random_state=42)

    scores = cross_val_score(clf, wbc_dataset.data, wbc_dataset.target, cv=5)

    return 1 - np.mean(scores) # Minimize!


# Creating Scenario object
scenario = Scenario({"run_obj": "quality",
                     "runcount-limit": 200, # maximum function evaluations
                     "cs": cs, # configuration space
```

```
                        "deterministic": "true"
                    })


# SMAC object handles bayesian optimization loop
print("Please wait until optimization is finished")
smac = SMAC(scenario=scenario, rng=np.random.RandomState(42),
        tae_runner=xgboost_from_cfg)

incumbent = smac.optimize()

# Let's see the best performing hyperparameter values
print(incumbent)
# Configuration:
# learning_rate, Value: 0.08815217130807515
# n_estimators, Value: 196

# You can see the errpr rate of optimized hyperparameters
inc_value = xgboost_from_cfg(incumbent)

print("Optimized Value: %.2f" % (inc_value))
# 0.02
```

非常好！现在你学会了如何创建配置空间、增加超参数以及为每个超参数定义取值范围。配置完成后，可以看到如何创建场景对象，使用 SMAC 优化已知估算器的超参数。

使用 SMAC 对象可以获取运行历史记录并查看每个配置的开销：

```
param_1 = []
param_2 = []
costs = []

for k,v in smac.runhistory.config_ids.items():
    param_1.append(k._values['learning_rate'])
    param_2.append(k._values['n_estimators'])
    costs.append(smac.runhistory.cost_per_config[v])

print(len(param_1), len(param_2), len(costs))

import matplotlib.pyplot as plt
import matplotlib.cm as cm

sc = plt.scatter(param_1, param_2, c=costs)
plt.colorbar(sc)
plt.show()
```

图 5-1 展示了 learning_rate 和 n_estimators 优化过程中的不同取值及相关开销。

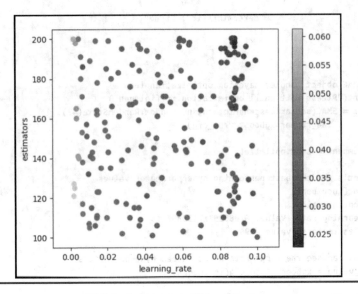

图 5-1　learning-rate 和 n_estimators

由此可见，learning_rate 最佳配置为~0.09，n_estimators 最佳配置为~200。

5.6　总结
Summary

本章中，我们学习了模型参数、超参数和配置空间参数。让我们快速回顾内容。

- **模型参数**：训练时可学习到的参数。
- **模型超参数**：训练开始前定义好的参数。
- **配置空间参数**：实验环境中用到的其他任意参数。

本章介绍了常用的超参数优化方法，如网格搜索和随机搜索。网格搜索和随机搜索未利用前一轮训练的信息，这一缺憾由贝叶斯优化法弥补。

贝叶斯优化法利用了前一轮训练的信息决定下一轮训练的超参数值，以更智能的方式遍历全部超参数空间。SMAC 是 auto-sklearn 内部用来为特定估算器进行超参数优化的方法。本章展示了如何在机器学习流水线中应用此方法。

第 6 章

创建 AutoML 流水线
Creating AutoML Pipelines

之前的章节关注的是完成一个机器学习项目所必经的不同阶段。若要成功运行机器学习模型并产出结果，需把各运行环节连接起来。将机器学习流程中不同环节连接在一起的过程叫构建流水线。流水线虽然是一个概括性概念，但是对数据科学家非常重要。在软件工程中，人们构建流水线进行软件开发，从源代码到部署全覆盖。相似地，在机器学习中，也会构建流水线，让数据流从原始格式转化为有用信息。构建多重机器学习并行流水线系统，将不同机器学习方法的结果进行对比。

流水线的每个阶段都会接受上一阶段处理过的数据，即当前处理单元的输出会成为下一步的输入。流水线中的数据流通如同管道中的水流一样。掌握流水线的理念有助于创建强大的无误差机器学习模型，而且流水线是构建 AutoML 的关键元素。本章中，关注主题如下。

- 机器学习流水线简介。
- 构建简单流水线。
- 函数转换器。
- 用弱学习器和集成学习器构建复杂流水线。

下面先从介绍机器学习流水线开始。

6.1 技术要求
Technical Requirements

本章的全部代码示例都可在 GitHub 链接的 Chapter 06 中找到。

6.2 机器学习流水线简介
An Introduction to Machine Learning Pipelines

一般来说,机器学习流水线需要干净的数据来检测数据中的某些模式,并在新的数据集中进行预测。但是,现实应用中的数据通常都无法直接用到机器学习算法中的。同样,机器学习的输出也只是数值或字符,要经过处理才可在实践中利用起来。要进行这样的处理,机器学习模型必须部署在生产环境中。从原数据转化到有用信息的整个框架均是由机器学习流水线完成的。

图 6-1 是对机器学习流水线的概括图示。

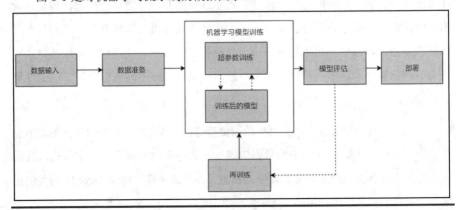

图 6-1 机器学习流水线

下面我们拆分图 6-1 中所示模块单独说明。

数据输入(data ingestion):为了便于使用而进行的数据获取和数据导入的过程。数据可源于多种系统,如 ERP(Enterprise resource planning,企业资产计划)软件、CRM(customer relationship management,客户关系管理)软件和 Web 应用。

数据可以是实时提取的，也可以是分批提取的。有时数据的获取比较棘手，这也是最具挑战的步骤之一，因为要求能很好地理解业务和数据。

数据准备（data preparation）：在第 3 章学习过各种数据预处理的技术。有许多方法可将数据预处理成恰当的形式，用于构建模型。现实中的数据常常是不完美的，有时缺失数据，有时有干扰项。因此，很有必要对数据进行预处理，包括清洗和转换数据，让数据在机器学习算法中能运行。

机器学习模型训练（ML model training）：理解数据中的必要特征、进行预测或从中衍生出独立的见解，就涉及使用各种机器学习技术。通常，机器学习算法的代码已经完成，并有 API 或程序接口可供调用。最重要的工作是超参数优化。超参数的使用和优化，从而构建最合适的模型，是模型训练阶段最关键、也是最复杂的环节。

模型评估（model evaluation）：评估模型的标准有许多。评估标准结合了统计方法和商业规则。在 AutoML 流水线中，评估大多是基于各种统计学和数学方法进行的。如果 AutoML 系统是为某商业领域或用例开发的，也要将其相应商业规则嵌入系统中来评估模型的准确性。

再训练（retraining）：为某个具体用例创建的第一个模型往往都不是最好的。可把第一个模型看成基线模型，反复训练以提升模型的准确性。

部署（deployment）：最后一步是模型部署，把模型应用并迁移到商业运营环境中使用。部署阶段高度依赖组织机构的 IT 基础设施和软件能力。

由此可见，从机器学习模型中收获成果需经历几个阶段。scikit-learn 提供了流水线功能，可以创建复杂的流水线。构建 AutoML 系统时，流水线会非常复杂，因为要涵盖许多不同的场景。但如果了解如何预处理数据并利用机器学习算法和应用各种评估指标，那么流水线就只是给这些碎片确定形状的问题。

使用 scikit-learn 设计一个非常简单的流水线。

6.3 简单的流水线
A simple Pipeline

首先，导入一个称为 iris 的数据集，这是 scikit-learn 样本数据集库中才有的（http://scikit-learn.org/stable/auto_examples/datasets/plot_iris_dataset.html）。数据集包含 4 个特征和 150 行数据。在流水线中执行以下步骤，使用 iris 数据集训练模型。问题陈述是使用 4 个不同的特征预测 iris 数据中的品种，如图 6-2 所示。

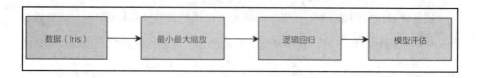

图 6-2 简单流水线

在这个流水线中，使用 MinMaxScaler 方法将来缩放输入数据，使用逻辑回归预测 iris 的种类。然后，评估模型准确率。

（1）从 scikit-learn 中导入各种库，提供完成任务所需的方法。在此前章节中，我们已经学习过这部分内容。唯一新增的是从 sklearn.pipeline 中导入 Pipeline，其中包含创建机器学习流水线的必要方法：

```
from sklearn.datasets import load_iris
from sklearn.preprocessing import MinMaxScaler
from sklearn.linear_model import LogisticRegression
from sklearn.model_selection import train_test_split
from sklearn.pipeline import Pipeline
```

（2）加载 iris 数据，分成训练数据集和测试数据集。本例中，使用 80%的数据训练模型，剩下的 20%用于测试模型准确率。使用 shape 函数查看数据集的维度：

```
# Load and split the data
iris = load_iris()
X_train, X_test, y_train, y_test = train_test_split(iris.data,
    iris.target, test_size=0.2, random_state=42)
X_train.shape
```

（3）以下结果展示了训练数据集有 4 列和 120 行，相当于 80%的 iris 数据集，与预期相符：

```
Out[22]: (120, 4)
```

（4）打印数据集，查看数据：

print(X_train)

以上代码输出如下：

```
[[4.6 3.6 1.  0.2]
 [5.7 4.4 1.5 0.4]
 [6.7 3.1 4.4 1.4]
 [4.8 3.4 1.6 0.2]
 [4.4 3.2 1.3 0.2]
 [6.3 2.5 5.  1.9]
 [6.4 3.2 4.5 1.5]
 [5.2 3.5 1.5 0.2]
 [5.  3.6 1.4 0.2]
 [5.2 4.1 1.5 0.1]
 [5.8 2.7 5.1 1.9]
 [6.  3.4 4.5 1.6]
 [6.7 3.1 4.7 1.5]
 [5.4 3.9 1.3 0.4]
 [5.4 3.7 1.5 0.2]
```

（5）创建流水线。流水线对象的形式是键-值（key，value）对。键是步骤名称字符串，值是函数或实际方法的名称。以下代码将 MinMaxScaler() 方法命名为 minmax，将 LogisticRegression(random_state=42) 命名为 lr：

pipe_lr = Pipeline([('minmax', MinMaxScaler()),
 ('lr', LogisticRegression(random_state=42))])

（6）将流水线对象 pipe_lr 拟合到训练数据集中：

pipe_lr.fit(X_train, y_train)

（7）以上代码输出如下，展示出了所构建拟合模型的最终结构：

```
Out[25]: Pipeline(memory=None,
        steps=[('minmax', MinMaxScaler(copy=True, feature_range=(0, 1))), ('clf', LogisticRegression(C=1.0, class_weight
    =None, dual=False, fit_intercept=True,
        intercept_scaling=1, max_iter=100, multi_class='ovr', n_jobs=1,
        penalty='l2', random_state=42, solver='liblinear', tol=0.0001,
        verbose=0, warm_start=False))])
```

（8）在 test 数据集中使用 score 方法给模型打分：

score = pipe_lr.score(X_test, y_test)
print('Logistic Regression pipeline test accuracy: %.3f' % score)

从以下结果可以发现，模型的准确率为 0.900，即 90%：

```
Logistic Regression pipeline test accuracy: 0.900
```

上述示例中，创建了一个流水线，由两步组成，分别是 minmax 缩放和 LogisticRegression。在 pipe_lr 流水线上执行 fit 方法，MinMaxScaler 在输入的数据中执行了 fit 和 transform 的方法，然后传送到估算器即逻辑回归模型。流水线中这些中间步骤叫**转换器（transformer）**，最后一步是估算器。

转换器用于数据预处理，有两种方法，fit 和 transform。fit 方法用于从训练数据中找出参数，transform 方法用于把数据预处理技术应用到数据集中。

估算器用于创建机器学习模型，有两种方法，fit 和 predict。fit 方法用于训练机器学习模型，而 predict 方法用于把训练过的模型应用到测试或新数据集中。

图 6-3 对概念进行了总结。

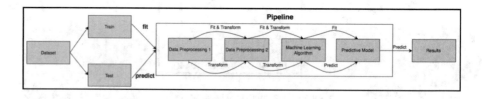

图 6-3　机器学习流水线概念

我们只需要调用流水线的 fit 方法就可以训练模型，调用 predict 方法就可以进行预测。其余的函数，即 Fit 和 Transform，是打包在流水线功能中的，按图方式执行。

有时，我们需编写一些自定义转换函数。下面讲述函数转换器，辅助实现自定义功能。

6.4　函数转换器 FunctionTransformer

FunctionTransformer 用于确定用户定义的函数，消费流水线中的数据，把函数结果返回给流水线的下一阶段。这是无状态转换，如取数值的平方或对数，确定自定义缩放函数等。

下例中，我们使用 CustomLog 函数和预定义的预处理方法 StandardScaler 来构建流水线。

（1）导入所要求的全部库，同上一个示例。唯一新增的是从 sklearn.preprocessing 库中增加 FunctionTransformer 方法。使用此方法执行自定义转换器函数，拼接到一起送入流水线其他阶段：

```
import numpy as np
from sklearn.datasets import load_iris
from sklearn.model_selection import train_test_split
from sklearn import preprocessing
from sklearn.pipeline import make_pipeline
from sklearn.preprocessing import FunctionTransformer
from sklearn.preprocessing import StandardScaler
```

（2）以下代码中，定义一个自定义函数，返回 X 值的 log 值：

```
def CustomLog(X):
    return np.log(X)
```

（3）现在定义一个叫 PreprocData 的数据预处理函数，接受数据集的输入数据（X）和目标值（Y）。本例中，Y 不是必需的，因为这次不是构建监督模型，只是演示数据预处理流水线。而在实践中，可直接使用这个函数创建监督机器学习模型。

（4）使用 make_pipeline 函数创建流水线。在前面的示例中使用过 pipeline 函数，需要定义数据预处理或机器学习函数的名称。make_pipeline 函数的好处是自动生成函数名称或键：

```
def PreprocData(X, Y):
pipe = make_pipeline(
 FunctionTransformer(CustomLog),StandardScaler()
 )
 X_train, X_test, Y_train, Y_test = train_test_split(X, Y)
 pipe.fit(X_train, Y_train)
 return pipe.transform(X_test), Y_test
```

（5）流水线准备就绪，可以加载 iris 数据集了。打印输入数据 X，查看数据：

```
iris = load_iris()
X, Y = iris.data, iris.target
print(X)
```

以上代码打印输出如下：

```
[[5.1 3.5 1.4 0.2]
 [4.9 3.  1.4 0.2]
 [4.7 3.2 1.3 0.2]
 [4.6 3.1 1.5 0.2]
 [5.  3.6 1.4 0.2]
 [5.4 3.9 1.7 0.4]
 [4.6 3.4 1.4 0.3]
 [5.  3.4 1.5 0.2]
 [4.4 2.9 1.4 0.2]
 [4.9 3.1 1.5 0.1]
 [5.4 3.7 1.5 0.2]
 [4.8 3.4 1.6 0.2]
 [4.8 3.  1.4 0.1]
 [4.3 3.  1.1 0.1]
 [5.8 4.  1.2 0.2]
 [5.7 4.4 1.5 0.4]
 [5.4 3.9 1.3 0.4]
 [5.1 3.5 1.4 0.3]
 [5.7 3.8 1.7 0.3]
```

（6）传入 iris 数据，调用 PreprocData 函数。结果返回转换后的数据集，先使用 CustomLog 函数处理，然后使用 StandardScaler 方法处理：

```
X_transformed, Y_transformed = PreprocData(X, Y)
print(X_transformed)
```

（7）以上的数据转换任务产生如下转换后的数据：

```
[[-0.2347759  -0.48960713  0.69581726  0.88662043]
 [ 0.00396322 -0.48960713  0.76352343  1.07005764]
 [ 0.34706281 -0.48960713  0.62528921  0.372669  ]
 [ 0.56654528 -0.25197392  0.92180958  0.7806152 ]
 [ 1.08613866  0.41464764  1.01013024  1.0272376 ]
 [ 1.18545964  0.19965017  0.76352343  1.0272376 ]
 [-0.11435918 -0.48960713  0.39414489  0.4532015 ]
 [ 0.9853462   0.19965017  0.51366007  0.52776295]
 [-0.48219832  0.82518858 -1.30762888 -0.73266584]
 [-0.3573624   1.02148788 -1.54981681 -1.43005448]
 [ 0.56654528 -0.73588342  0.69581726  0.7806152 ]
 [-0.7389596   0.82518858 -1.42439437 -1.43005448]
 [ 1.08613866 -0.48960713  0.66092059  0.52776295]
 [ 0.67368713  0.41464764  0.82862485  1.0272376 ]
 [-1.0057925  -1.82170004  0.02677876  0.18923179]
 [-1.0057925   0.82518858 -1.30762888 -1.43005448]
 [ 0.2346079  -0.02239764  0.66092059  0.7806152 ]
 [ 1.18545964  0.19965017  0.86025995  0.93570915]
 [ 0.67368713 -0.48960713  0.92180958  0.98251383]
 [ 0.12026292 -0.02239764  0.76352343  0.7806152 ]
 [-0.11435918 -0.02239764  0.43492827  0.372669  ]
 [ 0.00396322  1.92574265 -1.68528336 -1.43005448]
```

现在要创建各种复杂的 AutoML 系统流水线了。下面我们会使用几种数据预处理方法和机器学习算法创建一个复杂流水线。

6.5 复杂流水线
A Complex Pipeline

本节中，我们要使用 4 个不同的特征确定预测鸢尾花（Iris）花品种的最佳分类器。本任务将结合 4 种不同的数据预处理技术与 4 种不同的机器学习算法。图 6-4 是本任务的流水线设计。

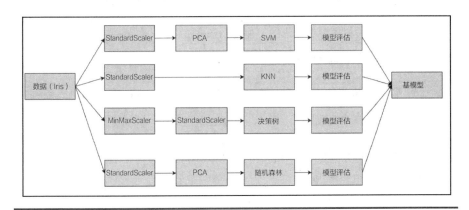

图 6-4　复杂流水线

创建复杂流水线的步骤如下。

（1）导入任务所需的各种库和函数：

```
from sklearn.datasets import load_iris
from sklearn.preprocessing import StandardScaler
from sklearn.decomposition import PCA
from sklearn.preprocessing import MinMaxScaler
from sklearn.model_selection import train_test_split
from sklearn.neighbors import KNeighborsClassifier
from sklearn.ensemble import RandomForestClassifier
from sklearn import svm
from sklearn import tree
from sklearn.pipeline import Pipeline
```

（2）加载 iris 数据集，分成 train 和 test 数据集。我们会用 X_train 和 Y_train

数据集训练不同的模型,并且用 X_test 和 Y_test 数据集测试所训练的模型:

```
# Load and split the data
iris = load_iris()
X_train, X_test, y_train, y_test = train_test_split(iris.data,
    iris.target, test_size=0.2, random_state=42)
```

(3)创建 4 个流水线,每个模型 1 个。在 SVM 模型流水线 pipe_svm 中,首先使用 StandardScaler 缩放输入数据,然后使用 PCA 创建主成分,最后使用预处理过的数据集构建 SVM 模型。

(4)相似地,构建名为 pipe_knn 的 KNN 模型流水线。只使用 StandardScaler 对数据进行预处理,然后执行 KNeighborsClassifier 构建 KNN 模型。

(5)创建决策树模型的流水线。使用 StandardScaler 和 MinMaxScaler 方法预处理用于 DecisionTreeClassifier 方法的数据。

(6)创建随机森林模型流水线。只使用 StandardScaler 预处理用于 RandomForestClassifier 方法的数据。

以下是创建 4 种不同模型流水线的代码:

```
# Construct svm pipeline

pipe_svm = Pipeline([('ss1', StandardScaler()),
      ('pca', PCA(n_components=2)),
      ('svm', svm.SVC(random_state=42))])
# Construct knn pipeline
pipe_knn = Pipeline([('ss2', StandardScaler()),
      ('knn', KNeighborsClassifier(n_neighbors=6,
metric='euclidean'))])

# Construct DT pipeline
pipe_dt = Pipeline([('ss3', StandardScaler()),
      ('minmax', MinMaxScaler()),
      ('dt', tree.DecisionTreeClassifier(random_state=42))])

# Construct Random Forest pipeline
num_trees = 100
max_features = 1
pipe_rf = Pipeline([('ss4', StandardScaler()),
      ('pca', PCA(n_components=2)),
      ('rf', RandomForestClassifier(n_estimators=num_trees,
max_features=max_features))])
```

（7）把流水线的名称存入字典中，用于展示结果：

```
pipe_dic = {0: 'K Nearest Neighbours', 1: 'Decision Tree',
2:'Random Forest', 3:'Support Vector Machines'}
```

（8）列出全部 4 个流水线，并依次执行：

```
pipelines = [pipe_knn, pipe_dt,pipe_rf,pipe_svm]
```

（9）现在，整个流水线的复杂结构已准备好。只剩下将数据拟合到流水线中，评估结果和选择最佳模型。

以下代码中依次将 4 个流水线拟合到训练数据集中：

```
# Fit the pipelines
for pipe in pipelines:
    pipe.fit(X_train, y_train)
```

（10）模型拟合执行成功后，使用以下代码检验 4 个模型的准确率：

```
# Compare accuracies
for idx, val in enumerate(pipelines):
    print('%s pipeline test accuracy: %.3f' % (pipe_dic[idx],
val.score(X_test, y_test)))
```

（11）从以下结果可见，KNN 和决策权模型以 100%的准确率领先。这个结果好到不真实，可能是因为数据集数量小或过度拟合导致的：

```
K Nearest Neighbours pipeline test accuracy: 1.00
Decision Tree pipeline test accuracy: 1.00
Random Forest pipeline test accuracy: 0.90
Support Vector Machines pipeline test accuracy: 0.90
```

（12）部署任意一个优秀模型，KNN 或决策树模型。使用以下代码可以完成：

```
best_accuracy = 0
best_classifier = 0
best_pipeline = ''
for idx, val in enumerate(pipelines):
 if val.score(X_test, y_test) > best_accuracy:
  best_accuracy = val.score(X_test, y_test)
  best_pipeline = val
  best_classifier = idx
print('%s Classifier has the best accuracy of %.2f' %
(pipe_dic[best_classifier],best_accuracy))
```

（4）由于 KNN 和决策树的准确率相同，可以就选 KNN 为最佳模型，因此，它是流水线中的第一个模型。但是，在这个阶段，也可采用一些商业规则或运行成本来决定哪一个是最佳模型。

```
K Nearest Neighbours is the classifier has the best accuracy of 1.00
```

6.6 总结
Summary

本章是粗略概览机器学习系统创建流水线的过程，这只是冰山一角。流水线创建非常复杂，但是，流水线研发成功后，后面的日子就舒服多了。流水线可以降低构建不同模型的复杂度，因而是创建 AutoML 系统所必不可少的。本章描述的一些概念为创建流水线奠定了基础。有了流水线后，模型构建流程会变得有条不紊。

第 7 章是对到目前为止所学到的全部内容进行总结。同样也会针对成功设计 AutoML 系统和开展数据科学项目提供一些有用的建议。

第 7 章

深度学习探究
Dive into Deep Learning

人工智能（AI）的未来重点在于自动化。本书已经覆盖了**自动机器学习**（AutoML）的基本知识。人工智能的另外一个领域在许多场景中刚刚崭露头角，也是亟需自动化实现的一个领域。这个领域就是**深度学习**（deep learning，DL）。深度学习是机器能力的顶尖体现，它比机器学习做得更多、更轻松且更准确。深度学习算法可以自己学习到数据集的关键特征，可调节权重，创建更好的模型等。深度学习的应用面非常广。

由于有了深度学习的先进技术，图像、语音和视频识别领域的学者和从业者看到了越来越多的实用成果。深度学习促进 AI 更接近其最初的目标：成为机器人的大脑。深度学习在不断增长的物联网中也发挥了作用。对企业而言，深度学习已经在简化客户服务和人力密集型任务自动化中做出了非常重要的贡献。目前你很可能已经在商品咨询或订皮萨的过程中体验过由深度学习驱动的机器客服系统。

深度学习在医疗和保健领域中有重大的发展空间。机器读取 X 射线和 MRI 核磁扫描进行诊断也不是新鲜事了。预计深度学习可解决许多人类的谜团，实现很多以前基本不可能的人工任务的自动化。有意思吧？现在我们一起解开深度学习的奥秘吧。

本章中，我们会学习以下主题。

- 前馈神经网络（feed-forward neural networks）。
- 自编码器（autoencoders）。
- 深度卷积神经网络（deep convolutional networks）。

本章中，我们将举例说明深度学习的一些基本概念。我们的重点不是用等式、数学公式和导数等让你从深度学习分神。虽然这些是核心概念，但有点太难理解。因此，我们会以演示的方式一点点接触这些概念，引导你写一些初级的深度学习代码。

神经网络是深度学习网络的先导，是构建深度学习框架的基石。神经网络背后的基本理念是创建一套基于我们生物大脑工作机制建模的计算系统。

7.1 技术要求
Technical Requirements

本章中的全部代码示例均可在 GitHub 的 Chapter 07 中找到。

7.2 神经网络概览
Overview of Neural Networks

神经网络的最佳定义是由最早的神经电脑科学家之一 Dr. Robert Hecht-Nielsen 在《神经网络入门：第一部分》（*Neural Network Primer: Part I*，作者为 AI 专家 Maureen Caudill，出版时间为 1989 年 2 月）中提出的：

> ……是由一些简单却高度互联的处理单元组成的计算系统，信息处理是通过这些单元对外部输入的动态响应完成的。

神经网络的基础单元是神经元，位于不同的分层结构中。每一层的神经元都与下一层的神经元相关联。神经网络至少有三层。输入层与一层或多层隐藏层相连，连接是通过加权链接构建的。最后的隐藏层与生成任务结果的输出层相连。

图 7-1 展示了一个三层神经网络。

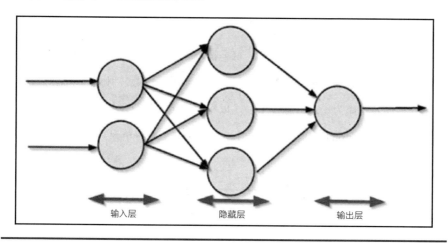

图 7-1 三层神经网络

此网络也叫全连接人工智能神经网络或**前馈神经网络**(feed-forward neural network, FNN)。另一种神经网络架构, 把结果传回进行再学习, 调节权重, 称为**带反向传播**(backpropagation)**的前馈神经网络**, 如图 7-2 所示。

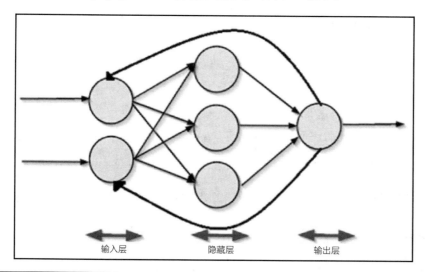

图 7-2 带反向传播的前馈神经网络

下面我们全面了解神经网络的每个单元。

神经元
Neuron

神经元是神经网络的基本单元。每个神经元有多个输入、一个处理器和一个输出。神经元的流程从聚集全部输入开始，基于激活函数，触发一个信号，传递到其他神经元或作为输出信号传出去，以帮助神经元学习。神经元最基本的形式是感知器（perceptron）。感知器是一个0或1的二值输出单元。

典型的神经元如图7-3所示。

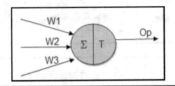

图7-3 典型的神经元

由图7-3可见，神经元有多个输入。每个输入连接都有与该连接关联的权重。神经元激活时，对几个输入与其对应连接权重的乘积求和，计算得出状态。数学上可表示为如下函数：

$$\sum_{i=1}^{m} w_i x_i$$

其中，x_i是输入值，包括偏置。

偏置是神经元的另外一个输入。偏置有自己的连接权重，偏置值的权重始终是1。偏置的作用是保证即使没有输入，即输入值为0时，神经元也有激活的机会。

状态计算出来后，传入激活函数，将结果归一化（0或1）。

激活函数
Activation Functions

激活函数使用加权输入的先验线性组合来生成非线性决策。这里将讨论深度学习应用中最主流的四种激活函数。

- step。
- sigmoid。
- ReLU。
- tanh。

step 函数

在 step 函数中,如果输入值权重之和超过某个阈值,神经元就会激活。两个选项如下:

$$\varnothing(x) = \begin{cases} 1, if x \geq 0 \\ 0, if x < 0 \end{cases}$$

如果输入值的权重和 x 大于或等于 0,则激活 1;否则激活 0。图 7-4 展示了 step 函数。

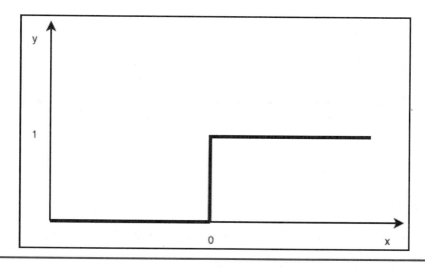

图 7-4　step 函数

下一个常用的激活函数是 sigmoid 函数。

sigmoid 函数

sigmoid 函数定义如下:

$$\emptyset(x) = \frac{1}{1+e^{-x}}$$

其中，x 是输入值的权重和。在逻辑回归中已经见过这个函数。当 x 小于 0 时，它就会无限接近于 0；大于 0 时，则无限接近于 1。它的作用是将正无穷与负无穷之间的区间压缩到（0，1）之间，与 step 函数不同，这是一个非线性激活函数。常用在神经网络中的输出层，用在试图预测概率的分类任务中。图 7-5 演示了 sigmoid 函数。

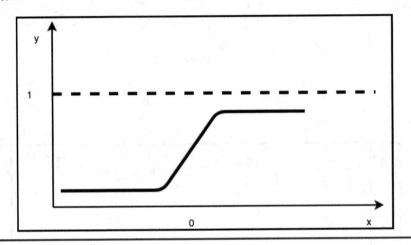

图 7-5 sigmoid 函数

下一个要讨论的激活函数是 ReLU，神经网络的隐藏层中经常用到此函数。

ReLU 函数

研究者发现，使用**修正线性单元**（rectified linear unit，ReLU）函数的神经网络比使用其他如 sigmoid 和 tanh 等非线性函数的神经网络训练得更快，同时准确率也没有明显下降。因此，ReLU 函数是最重要的激活函数之一。如果 x 是正数，则输出 x，否则输出 0。

定义如下：

$$A(x) = max(0,x)$$

ReLU 函数如图 7-6 所示。

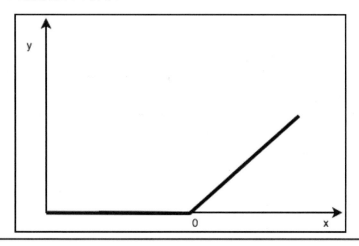

图 7-6　ReLU 函数

 ReLU 是非线性函数，ReLU 函数的组合也是非线性的。ReLU 的范围从 0 到无限大。

接下来讨论 tanh 函数，与 sigmoid 函数非常相似，但值可小于 0。

tanh 函数

tanh 函数也叫**双曲正切函数**（hyperbolic tangent function）。值从 −1 到 +1。

数学公式表示如下：

$$\varnothing(x) = \frac{1 - e^{-2x}}{1 + e^{-2x}}$$

tanh 函数输出是以 0 为中心的。这个函数也是非线性函数，因此可用于叠加不同的层。tanh 函数表示如图 7-7 所示。

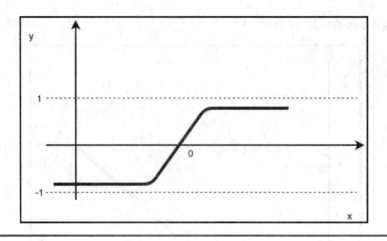

图 7-7　tanh 函数

现在对神经网络、神经网络结构及其不同组件都有了了解了,我们使用 Keras 创建一个前馈神经网络。

7.3　使用 Keras 的前馈神经网络
A Feed-forward Neural Network Using Keras

Keras 是最初使用 Python 创建的深度学习库,在 TensorFlow 或 Theano 上运行。开发这个库是为了让深度学习实现更快捷。

(1)在操作系统的命令行中使用以下代码调用 install keras:

```
pip install keras
```

(2)先导入 numpy 和 pandas 库,进行数据修改。同时,设置一个 seed,可以再现脚本结果:

```
import numpy as np
import pandas as pd
numpy.random.seed(8)
```

(3)分别从 keras.models 和 keras.layers 导入序贯模型和密度层。Keras 模型由一系列的层定义。序贯结构可以让用户配置和添加层。用户可利用密度层构建一个全连接的网络:

```
from keras.models import Sequential
from keras.layers import Dense
```

（4）加载人力流失数据集，有 14999 行和 10 列。对 salary 和 sales 属性进行独热编码，以便使用 Keras 构建深度学习模型：

```
#load hrdataset
hr_data = pd.read_csv('data/hr.csv', header=0)
# split into input (X) and output (Y) variables
data_trnsf = pd.get_dummies(hr_data, columns =['salary', 'sales'])
data_trnsf.columns
X = data_trnsf.drop('left', axis=1)
X.columns
```

以上代码输出如下：

```
Out[16]: Index(['satisfaction_level', 'last_evaluation', 'number_project',
                'average_montly_hours', 'time_spend_company', 'Work_accident',
                'promotion_last_5years', 'salary_high', 'salary_low', 'salary_medium',
                'sales_IT', 'sales_RandD', 'sales_accounting', 'sales_hr',
                'sales_management', 'sales_marketing', 'sales_product_mng',
                'sales_sales', 'sales_support', 'sales_technical'],
               dtype='object')
```

（5）按 70∶30 的比例将数据集分成两个子集，分别用于训练和测试模型：

```
from sklearn.model_selection import train_test_split

X_train, X_test, Y_train, Y_test = train_test_split(X,
data_trnsf.left, test_size=0.3, random_state=42)
```

（6）用三个层定义一个序贯模型，创建一个全连接神经网络。第一层是输入层。在输入层中，使用 input_dim 参数定义输入特征的数量，定义神经元的数量和激活函数。设置 input_dim 参数值为 20，预处理数据集 X_train 中有 20 个输入特征。第一层的神经元数量定为 12。我们的目标是预测员工流失，这是一个二分类问题。所以，要在前两层中使用 ReLU 激活函数，将非线性引进模型中，激活神经元。第二层是隐藏层，配置 10 个神经元。第三层是输出层，有一个神经元，带 sigmoid 激活函数，保证这个二分类任务的输出是 0 与 1 之中的其一：

```
# create model
model = Sequential()
model.add(Dense(12, input_dim=20, activation='relu'))
model.add(Dense(10, activation='relu'))
model.add(Dense(1, activation='sigmoid'))
```

（7）配置模型后，下一步是编译模型。编译过程中，指定 loss（损失）函数、

optimizer（优化器）和 metrics（指标）。由于要处理的是二分类问题，所以要把 loss 函数指定为 binary_crossentropy。指定 adam 为本练习使用的优化器。优化算法的选择对模型的良好性能很关键。adam 优化器是随机梯度下降算法的延伸，是一种广泛使用的优化器：

```
# Compile model
model.compile(loss='binary_crossentropy', optimizer='adam',
metrics=['accuracy'])
```

（8）使用 fit 方法把模型拟合到训练集中。使用 epoch 指定要向前和向后传递的数量。batch_size 参数用于声明每个 epoch 中要用到的训练样本数量。本例中，指定 epochs 数为 100，batch_size 为 10：

```
# Fit the model
X_train = np.array(X_train)
model.fit(X_train, Y_train, epochs=100, batch_size=10)
```

（9）执行以上代码，可看到进度，如下所示。当达到模型 fit 方法中指定的 epoch 数后，进程就会停止：

```
Epoch 1/100
10499/10499 [==============================] - 6s - loss: 0.6086 - acc: 0.7618
Epoch 2/100
10499/10499 [==============================] - 4s - loss: 0.5390 - acc: 0.7620
Epoch 3/100
10499/10499 [==============================] - 4s - loss: 0.5015 - acc: 0.7631
Epoch 4/100
10499/10499 [==============================] - 4s - loss: 0.4569 - acc: 0.7781
Epoch 5/100
10499/10499 [==============================] - 4s - loss: 0.3980 - acc: 0.8119
Epoch 6/100
10499/10499 [==============================] - 4s - loss: 0.3359 - acc: 0.8544
Epoch 7/100
10499/10499 [==============================] - 4s - loss: 0.3005 - acc: 0.8805
Epoch 8/100
10499/10499 [==============================] - 4s - loss: 0.2818 - acc: 0.8874
Epoch 9/100
10499/10499 [==============================] - 5s - loss: 0.2713 - acc: 0.8946
```

（10）评估模型。我们在评估指标中已经指定了准确率。最后一步中使用以下代码获得模型的准确率：

```
# evaluate the model
scores = model.evaluate(X_train, Y_train)
print("%s: %.4f%%" % (model.metrics_names[1], scores[1]*100))

X_test = np.array(X_test)
```

```
scores = model.evaluate(X_test, Y_test)
print("%s: %.4f%%" % (model.metrics_names[1], scores[1]*100))
```

以下结果显示模型准确率为 93.56%，测试准确率为 93.20%，很不错的结果。我们得到的结果可能与种子描述的结果略有不同：

```
 9024/10499 [========================>.....] - ETA: 0s
acc: 93.56%
 3072/4500 [====================>.........] - ETA: 0s
acc: 93.20%
```

 通过需要数据缩放才能从神经网络中取得好结果。

下面讨论自编码器，一种无监督深度学习技术，常用于非线性降维。

7.4 自编码器
Autoencoders

自编码器是一种可用于无监督深度学习技术，与之前学过的如**主成分分析**（PCA）之类的降维技术类似。但是，PCA 使用线性转换将数据从高维降到低维，但自编码器采用的是非线性转换。

自编码器由以下两部分组成。

- **编码器**（encoder）：编码器将输入压缩成较少数量的元素或位。输入被压缩到最大限度，也叫**潜在空间**（latent space）或**瓶颈**（bottleneck）。这些压缩的位也叫**编码位**（encoded bits）。

- **解码器**（decoder）：解码器基于编码位对输入进行重构。如果解码器可从编码位中重新反推出完全相同的输入，则可以说编码完美。但是，这只是理想场景，事实并非如此。重构误差是衡量解码器重构工作和判断自编码器准确率的一种方法。

对自编码器有了初步了解后，我们再从图 7-8 中视觉化地理解一下。

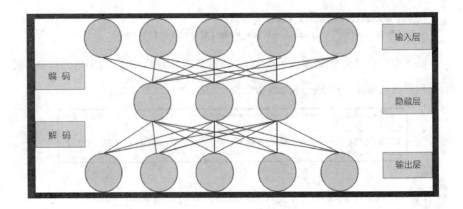

图 7-8 自编码器结构

自编码器有以下几种不同的类型。

- 传统自编码器（vanilla autoencoder）：两层神经网络架构，带一层隐藏层。
- 稀疏自编码器（sparse autoencoder）：用于学习数据的稀疏表示。对其损失函数加稀疏性限制以约束自编码器的重构。
- 去噪自编码器（denoising autoencoder）：在这些自编码器中引入部分损坏的输入。这么做是为了避免自编码器简单地学习输入的结构，强制神经网络发现更加可靠的特征来学习输入模式。

异常检测是自编码器中最广泛使用的用例之一。这是一个从未知数据中检测已知数据的过程。在异常检测练习中，输入数据往往会有一个类是**已知的**。自编码器重构输入数据，学习数据模式的时候，会更容易发现未知类，可能是数据中的异常值。

（1）同上一示例，使用以下代码导入所需库：

```
%matplotlib inline
import numpy as np
import pandas as pd
import matplotlib.pyplot as plt
from sklearn.model_selection import train_test_split
from keras.models import Model, load_model
from keras.layers import Input, Dense
np.random.seed(8)
```

（2）加载由自编码器实现的异常检测器用到的数据集。演示数据是从 1974 年 *Motor Trend US* 杂志中提取出来的，包括 32 辆汽车（1973—74 模型）的油耗和性能及汽车设计的 10 个方面。数据集被略加修改，引入了一些异常值：

```
# load autodesign
auto_data = pd.read_csv('data/auto_design.csv')
# split into input (X) and output (Y) variables
X =auto_data.drop('Unnamed: 0', 1)
from sklearn.model_selection import train_test_split
X_train, X_test = train_test_split(X, test_size=0.3, random_state=42)
print(X_train)
X_train.shape
```

以上代码输出如下：

	mpg	cyl	disp	hp	drat	wt	qsec	vs	am	gear	carb	ID
4	18.7	8	360.0	175	3.15	3.440	17.02	0	0	3	2	5
16	14.7	8	440.0	230	3.23	5.345	17.42	0	0	3	4	17
5	18.1	6	225.0	105	2.76	3.460	20.22	1	0	3	1	6
13	15.2	8	275.8	180	3.07	3.780	18.00	0	0	3	3	14
11	16.4	8	275.8	180	3.07	4.070	17.40	0	0	3	3	12
23	13.3	8	350.0	245	3.73	3.840	15.41	0	0	3	4	24
1	21.0	6	160.0	110	3.90	2.875	17.02	0	1	4	4	2
2	22.8	4	108.0	93	3.85	2.320	18.61	1	1	4	1	3
26	26.0	4	120.3	91	4.43	2.140	16.70	0	1	5	2	27
3	21.4	6	258.0	110	3.08	3.215	19.44	1	0	3	1	4
21	15.5	8	318.0	150	2.76	3.520	16.87	0	0	3	2	22
27	30.4	4	95.1	113	3.77	1.513	16.90	1	1	5	2	28
22	15.2	8	304.0	150	3.15	3.435	17.30	0	0	3	2	23
18	80.4	10	75.7	100	4.93	1.615	150.52	1	1	4	2	19
31	21.4	4	121.0	109	4.11	2.780	18.60	1	1	4	2	32
20	21.5	4	120.1	97	3.70	2.465	20.01	1	0	3	1	21
7	24.4	4	146.7	62	3.69	3.190	20.00	1	0	4	2	8
10	17.8	6	167.6	210	800.00	900.000	1000.00	1	0	4	4	11
14	10.4	8	472.0	205	2.93	5.250	17.98	0	0	3	4	15
28	15.8	8	351.0	264	4.22	3.170	14.50	0	1	5	4	29
19	33.9	4	71.1	65	4.22	1.835	19.90	1	1	4	1	20
6	14.3	8	360.0	245	3.21	3.570	15.84	0	0	3	4	7

（3）定义输入维度。由于有 12 个输入特征，而且会在自编码器中全部用到，所以定义输入神经元数量为 12。嵌入输入层，代码如下所示：

```
input_dim = X_train.shape[1]
encoding_dim = 12
input_layer = Input(shape=(input_dim, ))
```

（4）创建编码器和解码器。编码器中用到非线性激活函数 ReLU。编码层传入解码器中，重构输入数据模式：

```
encoded = Dense(encoding_dim, activation='relu')(input_layer)
decoded = Dense(12, activation='linear')(encoded)
```

（5）以下模型把输入映射到重构中，这一步是在解码层 decoded 中完成的。然后，使用 compile 方法定义 optimizer 和 loss 函数。adadelta 优化器使用指数衰减梯度平均值，是一种高学习率适应性方法。重构是线性过程，在解码器中使用线性激活函数定义的。loss 由 mse 定义，表示均方误差，在第 2 章中学习过：

```
autoencoder = Model(input_layer, decoded)
autoencoder.compile(optimizer='adadelta', loss='mse')
```

（6）拟合 X_train 训练数据到自编码器中。按 batch_size 为 4 训练 100 epoch 自编码器，观察是否达到稳定的训练值或测试损失值：

```
X_train = np.array(X_train)
autoencoder.fit(X_train, X_train,epochs=100,batch_size=4)
```

（7）执行以上代码，可看到进度，如下所示。达到 fit 方法中指定的 epoch 数（训练轮数）时，进程停止：

```
Epoch 1/100
22/22 [==============================] - 0s - loss: 28985.7149
Epoch 2/100
22/22 [==============================] - 0s - loss: 28571.5341
Epoch 3/100
22/22 [==============================] - 0s - loss: 28183.4400
Epoch 4/100
22/22 [==============================] - 0s - loss: 27798.2921
Epoch 5/100
22/22 [==============================] - 0s - loss: 27422.6434
Epoch 6/100
22/22 [==============================] - 0s - loss: 27080.4889
Epoch 7/100
22/22 [==============================] - 0s - loss: 26722.0817
Epoch 8/100
22/22 [==============================] - 0s - loss: 26391.1487
Epoch 9/100
22/22 [==============================] - 0s - loss: 26058.9774
Epoch 10/100
```

（8）模型拟合之后，将 X_train 数据集传递到自编码器的 predict 方法，预测输入值。然后，计算 mse 值，确定自编码器是否能够正确地重构数据集，重构误差有多大：

```
predictions = autoencoder.predict(X_train)
mse = np.mean(np.power(X_train - predictions, 2), axis=1)
```

（9）对 mse 绘图，查看重构误差以及无法正确重构的输入数据的索引：

```
plt.plot(mse)
```

（10）从图 7-9 可观察到，自编码器无法正确地重构数据集第 16 行记录。重构误差过大，该记录异常。第 13 条记录也有一个小重构误差，也可能是一个异常值。

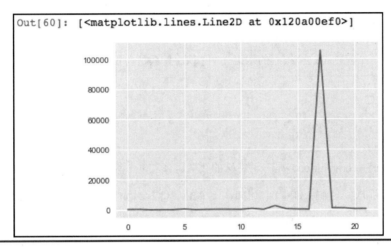

图 7-9　mse 图

下面重点介绍图像处理和图像识别中用得非常广泛的卷积神经网络（convolutional neural networks，CNN）。

7.5　卷积神经网络
Convolutional Neural Networks

本节重点介绍 CNN 架构。CNN 这个主题用一整章都未必能讲全，但在此着重挑选 CNN 的重要元素讲，让 CNN 的使用更容易一些。讨论过程中会用到 Keras 包，使用包里的 MNIST 样本数据集创建 CNN。

听到 CNN 这个词的时候，第一个出现在脑海的问题往往是为什么使用 CNN？这里简要地回答这个问题。

为什么使用 CNN
Why CNN?

前面已经讨论了前馈神经网络。FNN 已经很强大了，但一个主要的不足是忽略了输入数据的结构。所有输入给网络的数据首先都要转换成一维的数值数组。然而，比如在图像中，多维的数组就很难进行转换。保留图像结构是很有必要的，因为其中隐藏着大量信息，这时 CNN 就应运而生。CNN 处理图像的时候会考虑图像的结构。

下一个问题是这个很难的术语——卷积。什么是卷积？

什么是卷积
What is Convolution?

卷积（convolution）是一种特殊的数学运算，包含两个函数 f 和 g 的乘积，生成一个新的修正函数（f*g）。比如，一个图像（假设是函数 f）有一个过滤函数 g，两个函数的卷积就会生成一个新版的图像。

刚刚讨论了过滤器，那我们试着理解什么是过滤器。

什么是过滤器
What are Filters?

CNN 使用过滤器来识别图像中的特征，如边缘、线条或弧线，在图像中寻找某些重要模式或特征。过滤器设计用来搜索图像中的某些特性，检测图像中是否含有这些特性。过滤器应用在图像的不同位置中，直到整个图像都被过滤器覆盖。过滤器是卷积层中很关键的因素，是 CNN 的重要一步。

CNN 中主要有 4 个不同的层。

- 卷积层（convolution layer）。
- ReLU 层（ReLU layer）。

- 池化层（pooling layer）。
- 全连接层（fully connected layer）。

先讨论 CNN 的第一阶段：卷积层。

卷积层
The Convolution Layer

卷积层是在输入图像和刚讨论过的过滤器之间进行卷积运算的一层。这一步可以减小图像的总大小，后面各层中处理图像就会容易些。我们用一个简单的问题理解本层的功能：**如何识别一只狗或一只猫**？人们看到狗或猫的时候，自然瞬间就识别出来了。我们不需要分析全部的特征，确定狗是不是狗，猫是不是猫。

我们人能认出重要特征，如眼睛、耳朵或尾巴，然后就知道是什么物种。这也是卷积层要做的事情。在这一层中只识别重要特征，其他的全部忽略。在整个图像中移动过滤器，检测图像中的核心特征。移动过滤器的过程叫**步长**（strides）。

卷积层的结果会传到非线性激活函数中，如 ReLU 函数。

ReLU 层
The ReLU Layer

在卷积层中附加的一步是将非线性引入卷积特征图中。在前面章节已介绍过 ReLU 函数。图像具有高度非线性模式。应用卷积时，图像就会有变线性的风险，因为其中有乘法和加法等线性运算。所以，就使用如 ReLU 之类的非线性激活函数来保持图像中的非线性。

CNN 的下一阶段是池化层。

池化层
The Pooling Layer

池化层（pooling layer）通过应用池化函数来减小特征表示的大小。池化函数分为不同的类型，如 average（平均）、min（最小）和 max（最大）。最大池化用得很多，主要在每个步长中保留特征图的 max 值。与卷积层相似，池化层中有滑动窗口，沿着特征图滑动，在每个步长中找到 max 值。池化层的窗口大小一般比卷积层的要小。

池化特征图就会扁平化到用 1D（一维）表示，用在全连接层中。

全连接层
The Fully Connected Layer

一个 CNN 中可以有多个卷积、ReLU 和池化运算。但是，总会有一个最后阶段，叫全连接层。全连接层是我们前面讨论过的前馈神经网络。这一步的目的是对 image 数据集做出不同的预测，如进行图像分类。

图 7-10 展示了 CNN 基本架构的终极视图。

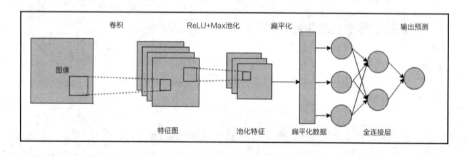

图 7-10　CNN 基本架构

理解了一些 CNN 基本知识后，现在就使用 Keras 创建一个 CNN。我们将使用 Keras 中已有的 MNIST 数据集。MNIST 数据集是有名的手写数字数据集。数据集已经分成训练集和练习集。其中约有 70000 张图片。每张灰度图尺寸为 28*28。

本节的完整代码可以在本书的代码库中找到。这里只展示一些重要的代码段。

从定义序贯模型开始构建 Keras 深度学习模型。

(1)如前面所讨论,序贯方法可以在其他层的基础上添加层,并按顺序执行。添加的第一层是由 Conv2D 方法定义的卷积层。由于 MNIST 数据集中包含的是 2D 图像,所以我们添加一个 2D 卷积层。使用 kernel_size 参数决定过滤器的大小,使用 strides 定义移动窗口的步长。

(2)Keras 中没有单独的 ReLU 层。但是,可在卷积层中定义激活函数。本次任务所选的激活函数为 ReLU。

(3)使用 MaxPooling2D 方法添加一个最大池化(max pooling)层。pool_size 定义为 2*2,池化层步长定义为 2*2。

(4)添加 Flatten2D 方法完成数据扁平化。最后一层是全连接层,与前馈神经网络的定义类似:

```
#Model Definition
cnn_model = Sequential()
cnn_model.add(Conv2D(32, kernel_size=(5, 5), strides=(1, 1),activation='relu',input_shape=input))
cnn_model.add(MaxPooling2D(pool_size=(2, 2), strides=(2, 2)))
cnn_model.add(Flatten())
cnn_model.add(Dense(1000, activation='relu'))
cnn_model.add(Dense(num_classes, activation='softmax'))
```

(5)以下代码段与所见过的其他 Keras 代码相似。首先使用 loss 函数、optimizer 和 metrics 编译模型。然后使用 batch_size 和 epochs 将模型拟合到训练集中,两个都是模型拟合方法中的重要参数:

```
cnn_model.compile(loss=keras.losses.categorical_crossentropy,
    optimizer=keras.optimizers.Adam(),
    metrics=['accuracy'])
cnn_model.fit(x_train, y_train,
    batch_size=10,
    epochs=10,
    verbose=1,
    validation_data=(x_test, y_test))
```

(6)执行以上代码后,要等待一段时间。只完成 10 个 epoch 也需要很长时间,因为训练数据集非常大而且图像又复杂:

```
Train on 60000 samples, validate on 10000 samples
Epoch 1/10
60000/60000 [==============================] - 566s - loss: 0.0994 - acc: 0.9697 - val_loss: 0.0685 - val_acc: 0.9779
Epoch 2/10
60000/60000 [==============================] - 533s - loss: 0.0406 - acc: 0.9879 - val_loss: 0.0501 - val_acc: 0.9835
Epoch 3/10
60000/60000 [==============================] - 554s - loss: 0.0240 - acc: 0.9925 - val_loss: 0.0451 - val_acc: 0.9877
Epoch 4/10
60000/60000 [==============================] - 577s - loss: 0.0172 - acc: 0.9945 - val_loss: 0.0551 - val_acc: 0.9872
Epoch 5/10
60000/60000 [==============================] - 558s - loss: 0.0152 - acc: 0.9960 - val_loss: 0.0580 - val_acc: 0.9877
Epoch 6/10
60000/60000 [==============================] - 692s - loss: 0.0121 - acc: 0.9968 - val_loss: 0.0787 - val_acc: 0.9851
Epoch 7/10
60000/60000 [==============================] - 59789s - loss: 0.0108 - acc: 0.9974 - val_loss: 0.0739 - val_acc: 0.9878
Epoch 8/10
60000/60000 [==============================] - 561s - loss: 0.0101 - acc: 0.9976 - val_loss: 0.0855 - val_acc: 0.9866
Epoch 9/10
60000/60000 [==============================] - 1463s - loss: 0.0083 - acc: 0.9979 - val_loss: 0.0983 - val_acc: 0.9855
Epoch 10/10
60000/60000 [==============================] - 537s - loss: 0.0098 - acc: 0.9980 - val_loss: 0.0924 - val_acc: 0.9875
<keras.callbacks.History at 0x11c954898>
```

（7）训练完成后，可使用以下代码评估模型的准确率：

```
model_score = cnn_model.evaluate(x_test, y_test, verbose=0)
print('Loss on the test data:', model_score[0])
print('Model accuracy on the test data:', model_score[1])
```

这个模型在测试数据集中的准确率很高，达到 0.9875，即 98.75%：

```
Loss on the test data: 0.09238649257670754
Model accuracy on the test data: 0.9875
```

7.6 总结
Summary

本章中，我们探索了神经网络和深度学习的世界。讨论了不同的激活函数、神经网络结构，并使用 Keras 演示了前馈神经网络的结构。

深度学习本身就是一个热点主题，有不少有关深度学习的专业书籍。本章的目的是为探索深度学习做一个开端，因为这是机器学习自动化的下一个前沿。我们见证了自编码器的降维能力。同时，也了解了 CNN 强健的特征处理机制，是构建未来自动化系统的必要组件。

第 8 章我们会结束学习，并回顾所学内容，介绍后续步骤，以及创建一套完整机器学习系统所需的技能。

第 8 章
机器学习和数据科学项目的重点
Critical Aspects of ML and Data Science Projects

看到这里，可以给自己点个赞了。虽不能自此就成为机器学习专家，但要鼓励自己为学习 AutoML 工作流程而付出的努力。现在，你可以应用这些技术解决自己的问题了！

本章中，我们会回顾各章节所学到的全部内容，并扩大学习范围。

讨论中会覆盖如下主题。

- 机器学习搜索。
- 机器学习中的权衡。
- 一个典型数据科学项目中的参与模型。
- 参与模型的各阶段。

8.1 机器学习搜索
Machine Learning As a Search

在前面所有章节中，看到许多技术应用于建模问题中，其中大多数技术虽然看似简单，但包含了很多能彻底影响结果的参数。许多建模都把 AutoML 变成了一个搜索问题，而且大多数情况下，只能搜索到一个次优解决方案。

更广义地看，建模只是输入数据和输出数据之间的映射。这样就可以根据新的输入数据来推断出未知的输出结果。为了实现这样的目标，就要考虑对应的实验设计和环境配置，因为你确实不清楚最好的机器学习流水线是什么。但我们先稍作暂停，向后退一步。

高性能系统的实现其实是从一些基础的系统架构选择开始的，选择好的系统架构，才有利于设计和交付成功的数据科学方案。首先开始考虑的事项之一应该是系统硬件和软件配置，如服务器类型、CPU 或 GPU 需求、内存和磁盘需求、软件要求等。如果要研究较大的数据集，配置就更重要了。这时的选择会决定数据科学堆栈的性能，可能会包含以下一些软件框架和库。

- 具体的软件发行包，如 Anaconda。
- 数据处理框架，如 Hadoop、Apache Spark、Kafka。
- 任务相关库，如 scikit-learn、XGBoost、TensorFlow/Keras、PyTorch。
- 数据管理系统，如 MongoDB、Neo4j、Apache Cassandra 和 MySQL。

这里列举得并不全，但即使在这个小范围内，要消化的信息也很多。理想情况下，你应该至少熟悉每种库在一个典型体系架构中所发挥的作用。这样在开始构建系统后，在实现不同用例系统的过程中，才会更加清楚如何选择。

软、硬件都就位且正常运转后，就要考虑数据流转问题，将数据输入机器学习流水线中。

在数据处理和建模阶段，选项十分广泛，如前所述，每种方法都有一套独立的参数。图 8-1 就是你已经练习过的典型机器学习工作流。

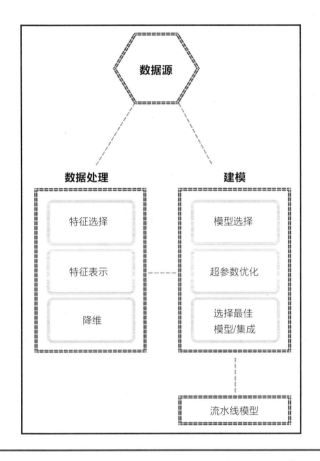

图 8-1 典型机器学习工作流

假设有一个文本处理流水线,用来预测财务市场活动,看看每一步在这种场景下表示什么。

首先,有各种数据源,可能来自以下途径。

- 金融交易所。
- 公司公告和文件。
- 提供综合和财经新闻的新闻机构。
- 政府机构宏观经济数据。

- 法规报告。
- 社交网络，如 Twitter 等。

图 8-2 展示了一些不同的数据源。

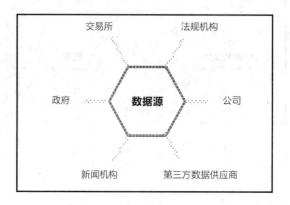

图 8-2 数据源

保存这些数据后，需要对数据进行处理，用于建模阶段，因为机器学习需要数值型向量输入。

第一步是特征选择，可能包含以下步骤。

- 确定是否每个词都作为不同的特征处理，这个过程通常称为**词袋处理**（bag-of-words）。
- 确定名词词语、动词词语或名词实体是否可用作特征。
- 将词语划分为代表性概念。
- 创建 n-grams，条目连续序列。

图 8-3 有助于理解特征选择。

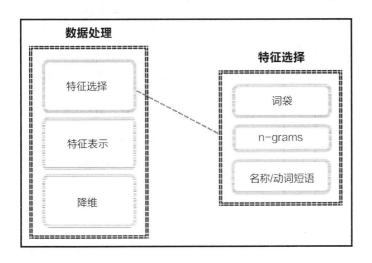

图 8-3 特征选择

确定了采用哪些特征，就要考虑减少维度，避免维度魔咒。这个阶段可能包含如下步骤。

- 简单的过滤操作，如只包含前 100 个概念。
- 设置阈值，只包含出现频率在已知阈值之上的词语。
- 使用各领域专家创建的预定义词典进行输入数据过滤。
- 标准化操作，如移除停止词和词形还原。

图 8-4 有助于了解特征选择和表示的不同方法。

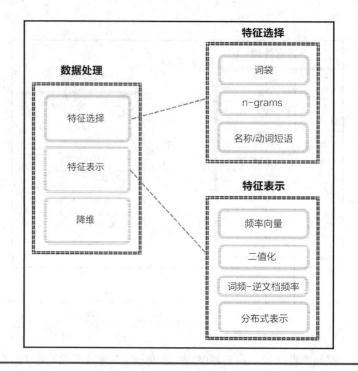

图 8-4 特征选择和表示

降维也是特征表示之后要返回的一个阶段。有了数值特征向量后，可应用如**主成分分析（PCA）**等方法进一步降维。特征表示的典型操作如下。

- 将词语表中的词语表示成频率向量，分配给每个词一个表示其出现次数的数字。

- 将频率向量二值化，每个值是 0 或 1。

- 用**词频-逆文档频率**（term frequency-inverse document frequency，tf-idf）编码，表示字词对于文档集即语料库（corpus）中的文档的重要性。

- 采用分布式表示，如 Word2Vec 或 Doc2Vec。

这样，图 8-5 描述了数据处理的完整流程。

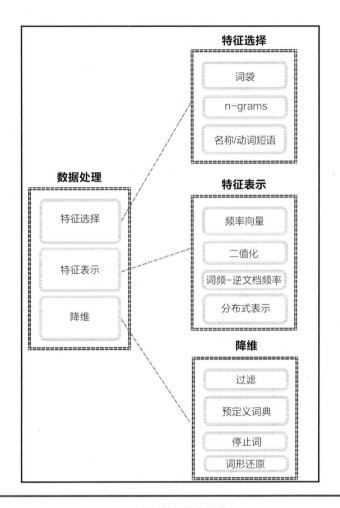

图 8-5　数据处理的完整流程

数据处理部分完成后，就可以开始建模了，本示例有许多不同的算法可选。以下是文本挖掘中最常用的算法。

- 深度神经网络，尤其是**长短期记忆**（Long Short-Term Memory，LSTM）网络，一种特殊的循环神经网络（recurrent neural networks，RNNs）类型。
- 支持向量机。
- 决策树。

- 集成。
- 回归。
- 朴素贝叶斯。

图 8-6 展示了几种可用的算法。

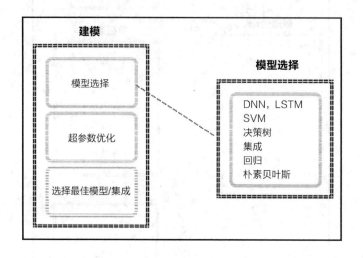

图 8-6 模型选择算法

每种算法都有自己的参数空间,参数主要包含以下两类。

- 超参数,训练开始前设置的参数。
- 模型参数,训练时学到的参数。

我们的目标是优化超参数,让模型参数产生最好的泛化能力。搜索空间一般非常大,我们已经接触过一些方法,比如贝叶斯优化,能以很高效的方式探索空间。

作为示例,我们看一下统领机器学习应用的 XGBoost 算法的参数。

运行以下代码,可看到模型参数说明:

```
import xgboost as xgb
classifier = xgb.XGBClassifier()

classifier?
```

输出如下：

```
Type: XGBClassifier
String form:
XGBClassifier(base_score=0.5, booster='gbtree', colsample_bylevel=1,
 colsample_bytree=1, g <...> reg_alpha=0, reg_lambda=1, scale_pos_weight=1, seed=None,
 silent=True, subsample=1)
File: ~/anaconda/lib/python3.6/site-packages/xgboost/sklearn.py
Docstring:
Implementation of the scikit-learn API for XGBoost classification.
 Parameters
 ----------
max_depth : int
  Maximum tree depth for base learners.
learning_rate : float
  Boosting learning rate (xgb's "eta")
n_estimators : int
  Number of boosted trees to fit.
silent : boolean
  Whether to print messages while running boosting.
objective : string or callable
  Specify the learning task and the corresponding learning objective or
  a custom objective function to be used (see note below).
booster: string
  Specify which booster to use: gbtree, gblinear or dart.
nthread : int
  Number of parallel threads used to run xgboost. (Deprecated, please use n_jobs)
n_jobs : int
  Number of parallel threads used to run xgboost. (replaces nthread)
gamma : float
  Minimum loss reduction required to make a further partition on a leaf node of the tree.
min_child_weight : int
  Minimum sum of instance weight(hessian) needed in a child.
max_delta_step : int
  Maximum delta step we allow each tree's weight estimation to be.
subsample : float
  Subsample ratio of the training instance.
colsample_bytree : float
  Subsample ratio of columns when constructing each tree.
```

```
colsample_bylevel : float
 Subsample ratio of columns for each split, in each level.
reg_alpha : float (xgb's alpha)
 L1 regularization term on weights
reg_lambda : float (xgb's lambda)
 L2 regularization term on weights
scale_pos_weight : float
 Balancing of positive and negative weights.
base_score:
 The initial prediction score of all instances, global bias.
seed : int
 Random number seed. (Deprecated, please use random_state)
random_state : int
 Random number seed. (replaces seed)
missing : float, optional
 Value in the data which needs to be present as a missing value. If
 None, defaults to np.nan.
**kwargs : dict, optional
 Keyword arguments for XGBoost Booster object. Full documentation of
 parameters can
 be found here:
 https://github.com/dmlc/xgboost/blob/master/doc/parameter.md.
 Attempting to set a parameter via the constructor args and **kwargs dict
 simultaneously
 will result in a TypeError.
 Note:
 **kwargs is unsupported by Sklearn. We do not guarantee that parameters
 passed via
 this argument will interact properly with Sklearn.
```

优化如 XGBoost 的超参数，需要选择一种具体的方法，如网格搜索、随机搜索、贝叶斯优化或者进化优化。实践中，贝叶斯优化的超参数优化结果通常较好。

图 8-7 所示为在流程里显示优化中常用的超参数和方法。

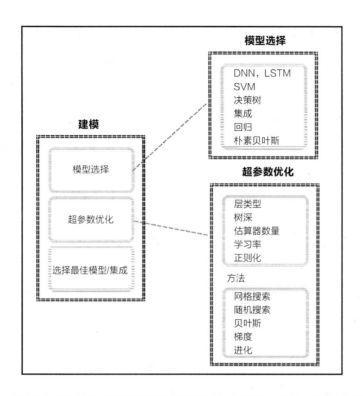

图 8-7 超参数优化

最后，选择最佳性能的机器学习流水线，或集成多个流水线。

掌握这些流水线需要对每一步都比较熟悉，要能合理浏览整个搜索空间，找到可接受的次优方案。

8.2 机器学习的权衡
Trade-offs in Machine Learning

机器学习主要有两方面要权衡。

- 训练时间。
- 推理时间。

这两个因素都会成为开发流水线时的限制条件。

想一想训练时间和推理时间可能的局限性。训练时间通常要求会影响候选算法的范围。比如，逻辑回归和**支持向量机**（SVM）是快速训练算法，这可能是重要因素，尤其是想用大数据快速将想法原型化的时候。这两种算法的推理也很快速。两种方法的实现不同，解决方案的选项也不同，这些解决方案能便捷地把两种算法应用到许多机器学习用例中。

但是，对于深度神经网络之类的算法，训练时间和推理时间是很大的限制条件，因为你可能没有办法容忍长达一周的训练时间或超过1秒的推理时间。选用性能更好的硬件资源可缩短训练时间和推理时间，但可能导致成本很高。网络越复杂，成本越高。除了算法选择，训练时间也高度依赖超参数空间，因此它可能导致所用时间更长。

另一个要考虑的因素是可扩展性，与数据量大小相关，需要保证流水线的可扩展性能匹配上数据增长的速度。这时，就应该看对多线程、多核、并行或分布式训练的支持情况等。

8.3 典型数据科学项目的参与模型
Engagement Model for a Typical Data Science Project

学习任何新知识或扩充已有知识，了解背景并理解故事发展史是很重要的，要明白当前的趋势是常规报表、商业智能（business intelligence，BI）和分析的自然演化结果。这也是我们在开始章节中要了解机器学习背景以及数据预处理、自动化算法选择和超参数优化等基本概念的原因。

由于AutoML流水线的高度试验性，所以有许多概念是结合实际示例解释的。

高级分析和机器学习用于解决问题的概念不一定都是新的，但是到现在才能用起来，因为硬件和软件资源的成本现在承担得起了。也有一些更高级的技术可解决以前解决不了的问题。

本书中，我们已经从各方面学习了如何开发机器学习流水线。但是在现实项目中，开发机器学习流水线只是变量之一。把机器学习流水线运转起来是极其重要的，因为只有成功部署和监控流水线后才能从优秀的机器学习流水线中受益。许多公司提供机器学习的 SaaS（Software as a Service，SaaS）服务，目标是让用户从管理生产环境机器学习流水线的低级复杂工作中解放出来。

下面将讨论数据科学项目的不同阶段。这样在端到端数据科学项目成功交付时，你能准确理解建模所处的阶段。

8.4 参与模型的阶段
The Phases of an Engagement Model

有一个非常知名的建模流程，该流程用于数据挖掘和预测分析项目的 CRISP-DM，包含以下六步。

- 业务理解。
- 数据理解。
- 数据准备。
- 建模。
- 评估。
- 部署。

一个阶段接着另一个阶段，还有一些会递归向前一阶段提供反馈。部署阶段的模型监控和维护尤其重要，这也是本章的重点。

我们快速浏览每一个阶段,并理解各阶段在整个过程中的作用。

业务理解
Business Understanding

这个阶段完全关注项目的目标、范围、资源、限制条件、迭代版本和检查点。

从业务角度理解整个故事,从技术角度搭建问题框架。组织内部不同利益相关方之间可能会有相左的目标,应该留意并找到折中方案。比如,供应链管理、生产供应部门想要将库存维持在最优级别,而销售部门又想根据非常乐观的销售预期为即将发布的产品建立大量库存。你应该知道谁会从即将实现的机器学习和数据科学能力中受益。

在此阶段,通常要努力解决以下问题。

- 了解当前决策过程,把不同的场景拆分成不同的用例。
- 约定人力资源,识别出要用的数据源(比如数据摘要、数据湖、运营数据库等)。
- 提出假设,尝试以可用数据来验证假设,分清观点/直觉和事实,再决定下一步行动。
- 约定交付物和整个项目中交付要用到的工具/技术。
- 确定机器学习模型和业务结果中要使用哪种评估指标/KPI。要始终与业务目标保持一致。
- 识别出可能导致延期或项目失败的风险,并将风险与相关者陈述清楚。机器学习项目的成功天然是概率性的,可不像大家习惯使用的商业智能系统(BI)那样保证成功。

数据理解
Data Understanding

这个阶段要理解在整个项目中用到的数据源。

本阶段中，通常要努力解决以下问题。

- 理清数据访问和授权问题。

- 将数据加载到一个首选的平台进行初步分析。

- 注意敏感信息，执行必要的操作，如敏感数据脱敏或删除。

- 识别要使用的数据集。

- 识别数据格式，获得字段描述。

- 确定每个数据集的数量，识别出数据集之间的差异。比如，检查不同的表格中是否存在具有不同数据类型的相同变量，如一个变量在一张表中是整形类型，在另外的表中是类别类型。

- 探索样本数据集，比如生成基本信息，包括数量、均值、标准方差、百分数等，并检查变量分布。

- 熟悉数据是如何收集的，在数据收集过程中是否可能有测量误差。

- 搞清楚关联关系，并铭记统计学谚语——**关联关系不代表因果关系**。

- 检测噪音干扰，如异常值，并约定如何处理异常值。

- 检查数据集的不同属性，确保样本数据集能代表总体。比如，是否会由于有些数据获取不到而偏向数据的某个类别。

- 快速反复检查质量，以保证数据质量。比如，找出并说明误差和缺失值，如日期列中有 100000000000，或均值为 23.4 的数据中出现 25000 等。

- 确认用于测试的数据量，从训练数据中预留。

最终，几次参与之后，这些实践就会引导你创建自动化数据预处理工作流，并生成数据质量报告。

数据准备
Data Preparation

这个阶段是通过组合不同的数据源、清洗、格式化和特征工程化而形成建模阶段要用到的最终数据集。

这一阶段通常要努力解决以下问题。

- 识别建模的关联数据集。
- 记录数据连接和聚合，构建最终数据集。
- 使用有效参数编写函数，便于在项目后期灵活清洗和格式化数据集，如移除 x% 的异常值，或使用均值、中值或最高频值填充缺失值。
- 酌情处理异常值。
- 尝试特征工程方法。
- 选择特征。总体而言，特征选择有以下三种主要方法：
    - 过滤法；
    - 包装法；
    - 嵌入法。
- 确定特征的重要性，列出包含/排除特征的原因。
- 约定要构建的数据转换流水线。
- 为具体操作编写自定义转换器，如通过扩展 Apache Spark 或 scikit-learn 的转换类从已知语料库中提取段落、句子和词语，类似地编写自定义估算器。
- 合理地标注数据集版本，添加备注，解释准备所需步骤。

建模
Modeling

本阶段为前面阶段所创建的最终数据集准备建模的选项。

本阶段中，通常要努力解决以下问题。

- 确定机器学习问题的类型，如监督、半监督、无监督或强化学习。

- 选出预算内的机器学习模型。

- 约定评估指标，注意重点，如分类失衡可能会影响准确率等指标。如果数据集不均衡，可尝试使用抽样技术得到平衡的数据集。

- 确定对假阴性和假阳性的容忍度。

- 考虑如何合理设置交叉验证。

- 分析性能模型的最重要特征。

- 针对每个特征分析模型敏感性。不同的算法可能对特征的排名不同；这有助于理解属性分布特征随时间变化时，模型会如何反应。

- 微调模型。

- 确定适合部署的机器学习负载类型，如在线、批量和流式部署。

- 考虑训练时间和推理时间，因为计算复杂度和数据大小通常是选择算法的限制因素。

评估
Evaluation

本阶段将回顾整个流程，以保证全部计划项目均已覆盖。

本阶段，通常要努力解决以下问题。

- 审查全部阶段，保证部署的模型流水线能解决全部问题，并检查此前进行的步骤。

- 准备一份精美的陈述，简洁但不简单。

- 呈现结果，切中要点，阐释模型如何满足业务目标。

- 说明限制条件，什么条件下会与目标相左。

部署
Deployment

这是开始运行机器学习模型的阶段。

本阶段，通常要努力解决以下问题。

- 简短的生产试运行。
- 监控性能。
- 找出性能改进策略，如模型性能开始衰减时，重新评估、重新训练、重新部署的策略。
- 准备最终报告和交付物。

每个阶段要解决的问题列表可一直追加下去，但本文提到的列表让你对可能要做哪些事情有一个总体的概念。

未来你可能有无数次从项目半途中加入其他项目组的机会。这种情况下，因为缺乏项目背景，所以需要与关键人物召开几次会议来了解进度。掌握了这些阶段，可以更好地理解项目当前的阶段，检查还有什么缺失，或搞清楚接下来的步骤。

经历这些步骤时，一定要记录每一步的原因和成果。完善的记录会在重复某些步骤时节约时间，因为随着时间的推移，可能会忘记怎么重现某些分析或数据处理的方法。

8.5 总结
Summary

如同我们一生所追寻的其他目标一样，唯有实践是提高开发机器学习流水线技能的方法。你应该投入大量时间学习不同的技术和算法来处理各种问题和数据集。

尤其在现实项目中，你可能不会遇到相似的问题，每个项目都要求不同的方法。很快就会发现不只建模重要，理解如何集成某些技术，在企业软件架构内表现出色才更重要。

学习 AutoML 系统，你又向前迈出了一大步，对 AutoML 流水线有了更深入的理解。当然还应该再继续努力学习，如你感兴趣领域中的具体应用、应用架构、生产环境和模型维护技术等。

感谢你付出时间学习本书内容，非常希望你和我们一样享受阅读本书！

作者简介
About the Authors

Sibanjan Das 是商业分析师和数据科学咨询师。他对预测性分析方案在商业系统和 IoT 领域的应用有丰富的经验。Sibanjan Das 对技术和创新满怀热情,在他职业生涯早期就积极投身于数据探索领域。Sibanjan Das 毕业于新加坡管理大学(Singapore Management University),IT 硕士,商业分析师,并拥有许多行业认证,包括 OCA(Oracle 认证工程师)、OCP(Oracle 认证专家)和 CSCMS(多元化内容管理系统专家)等。

本书得以完成,要感谢我的父亲 M. N. Das 医生、母亲、祖父母、姐妹、兄弟和连襟。同时,也感谢我的妻子以及儿子 Sachit,感谢他们对我的爱和关心。出版本书是我多年以来一直想做的事,是他们给了我动力。给我启发的人还有很多,在此无法一一列出。感谢所有人无尽的鼓励和善意的支持。

Umit Mert Cakmar 是 IBM 公司的数据科学家。他在 IBM 公司的工作是从可部署资产的构建到交付全面覆盖,帮助客户解决各种复杂的数据问题,表现不凡。除本行业研究外,Umit Mert Cakmar 还广泛涉猎其他领域,并多次在各行业大会、高校和见面会上分享自己的见解。

我要对我的父母致以最深切的谢意,谢谢他们对我的爱和支持。他们传授的经验让我受益匪浅。我想将本书献给我的家人、朋友、同事和所有坚持不懈地让世界变得更好的伟大人士。

索引
Index

A

activation functions
  about 220
  ReLU function 221
  sigmoid function 221
  step function 220
  tanh function 222
agglomerative clustering algorithm
  about 74
  implementing 165
Area under Curve (AUC) 52
artificial intelligence (AI) 6, 34, 216
auto-sklearn
  reference link 23
autoencoder
  about 227, 231
  decoder 227
  denoising autoencoder 228
  encoder 228
  sparse autoencoder 228
  vanilla autoencoder 228
automated feature preprocessing 18
Automated machine learning (AutoML)
  about 8, 9, 11, 15, 33, 216, 238
  automated algorithm selection 18
  automated feature preprocessing 18
  components, working 19
  consideration 17
  core components 17
  hyperparameter optimization 19
  learning 17
  prototype subsystems, building for component 19
  usage 16
Automated ML (AutoML) pipelines 196
AutoML libraries
  featuretools 20, 23

MLBox 25, 28, 30
overview 20
Tree-Based Pipeline Optimization Tool (TPOT) 30

B

bag-of-words 241
Bayesian optimization 181
Bayesian-based hyperparameter tuning 198
big O notation 127
binning
  about 102, 104
  equifrequency binning 103
  equiwidth binning 103
bootstrap aggregation 66
box and whisker 97
business intelligence (BI) 7, 248

C

CatBoost 151
categorical data transformation
  about 105
  encoding 105
  missing values 108, 109, 111
categorical feature creation 124
classification algorithms
  about 54
  confusion matrix 54
classifiers
  results, comparing 71
  URL 145
clustering
  about 73
  density-based technique 74
  grid-based method 74
  hierarchical clustering 74, 75

implementing 74
  partition-based clustering 74
  partitioning clustering (KMeans) 77
  using 73
code profiling
  in Python 132
coefficient of determination 47
complex pipeline 212, 215
computational complexity
  about 127
  Big O notation 127
configuration parameters 181
confusion matrix 54
Convolutional Neural Networks (CNN)
  about 231, 232
  convolution 232
  convolution layer 233
  features 232
  filters 232
  fully connected layer 234, 236
  pooling layer 234
  ReLU layer 233
corpus 243
cost function 181
CRISP-DM model
  URL 13
cross-validation 72

D

data transformation
  about 81
  categorical data transformation 105
  numerical data transformation 82
  text preprocessing 112, 113, 116
decision boundaries
  drawing 139
  of logistic regression 141
  of random forest 142
decision trees
  about 57
  implementing 57, 59
  using 57
Deep Feature Synthesis (DFS) 20
deep learning (DL) 216
Density-Based Spatial Clustering of Applications
  with Noise (DBSCAN) 153
  implementing 163
density-based technique 74
dimensionality reduction
  principal component analysis (PCA) 121
  using, for feature selection 121

E

encoded bits 227
encoding
  about 105, 106, 107
  binary encoding 106
  frequency-based encoding 106
  hash encoding 106
  label encoding 106
  one-hot encoding 106
  target mean encoding 106
engagement model
  business 249
  data 250
  data preparation 251
  deployment 253
  evaluation 253
  for data science project 248
  modeling 252
  phases 249
ensemble methods
  about 65
  ensemble models 65
ensemble models
  about 65
  bagging 66
  blending 71
  boosting 68, 71
  stacking 71

F

False Negative (FN) 55
false positive rate (FPR) 56
feature generation
  about 123
  categorical feature creation 124
  numerical feature generation 123
  temporal feature creation 124

feature selection
　about　116
　dimensionality reduction, using　121
　features, excluding with low variance　117
　randomForest, using　120
　recursive feature elimination　119
　univariate feature selection　118
feature transformations　144
features　34
featuretools
　URL　20
feed-forward neural network (FNN)
　about　218
　Keras, using　223
fitness function　181
function transformer　209, 211

G

Gradient Boosting Machines (GBMs)　69
grid-based method　74

H

hierarchical clustering　74, 75
hyperbolic tangent function　222
hyperparameters　10, 182, 184, 188, 194

I

Identifier detection　81
instance-based　130
inter-quartile range (IQR)　98
Internet of Things (IoT)　216

K

k-Nearest Neighbors (KNN)
　about　62, 63, 130
　implementing　63
　using　63
Keras
　using, for feed-forward neural network (FNN)　223
kernel trick　61
key performance indicators (KPIs)　7

L

label　10
latent space or bottleneck　227
lazy learners　130
LightGBM　151
linear regression
　about　36
　implementing　39, 43, 44, 45
　OLS regression, working　38
　OLS, assumptions　39
　using　39
linearity
　versus non-linearity　139
list wise deletion　88
Local Outlier Factor (LOF)　101
log and power transformation　104, 105
logistic regression
　about　48
　implementing　49, 52
　using　49
Long Short-Term Memory (LSTM)　44
loss function　11, 181

M

machine learning (ML)
　about　6, 33, 34, 80, 126, 181, 216, 238
　as search　238, 247
　process　34
　scope　7
　supervised learning　35
　trade-offs　247
　unsupervised learning　36
machine learning algorithms
　using　143
machine learning pipeline
　about　204
　Data Ingestion　204
　Data Preparation　205
　deployment　205
　ML model training　205
　Model Evaluation　205
　retraining　205
mean absolute error (MAE)　46
mean squared error (MSE)　44, 46

min-max normalization 85
MLBox
  reference link 25
model training 9
multiple linear regression 37

N

neural networks
  activation functions 220
  neuron 219
  overview 217
nominal 50
non-parametric 181
normalization 82
novel data 102
numerical data transformation
  about 82
  binning 102
  log and power transformation 104, 105
  missing values 87, 88, 93
  outliers 95
  scaling 82, 83, 84, 85, 86
numerical feature generation 123

O

one-hot encoding 50
ordinal 50
ordinary least square (OLS) 38
outliers
  about 95
  inter-quartile range 98
  multivariate outliers, detecting 101
  multivariate outliers, treating 101
  trimming 101
  univariate outliers, detecting 96
  univariate outliers, treating 96
  values, filtering 98
  winsorizing 100

P

pairwise feature creation 124
partitioning clustering (KMeans) 77
perceptron 219
predicators 34

predict method 208
predictors/features 35
Preprocessors
  URL 145
Principal Component Analysis (PCA)
  about 18, 121, 144, 169, 213, 227, 243
  implementing 171, 174

Q

Quantity Sold 37

R

R2 score 47
random forest model
  decision boundaries 142
  using, for feature selection 120
Receiver Operating Characteristic (ROC) 53
Rectified Linear Unit (ReLU) function 221
recurrent neural networks (RNNs) 244
recursive feature elimination (RFE) 119
regression algorithms, metrics
  mean absolute error (MAE) 46
  mean squared error 46
  R2 score 47
  root mean square error (RMSE) 47
regression algorithms
  about 46
  metrics 46
Regressors
  URL 145
Revenue 37
root mean square error (RMSE) 47

S

scaling 82
Sequential Model-based Algorithm Configuration (SMAC) 198
Sequential Model-based Global Optimization (SMBO) 198
sigmoid function 221
Silhouette Coefficient
  about 152
  cohesion 152
  separation 152

simple pipeline　206, 209
Software as a Service (SaaS)　16, 249
standardization　82
step function　220
Stochastic Gradient Boosting (SGB)　69
sum of squared errors (SSE)　38
supervised learning　35
Supervised ML
　about　145
　default configuration, of auto-sklearn　145
　pipeline, searching for network anomaly detection　151
　pipeline, searching for product line prediction　147, 151
Support Vector Machines (SVMs)
　about　15, 35, 60, 139, 213, 248
　implementing　61, 62
　using　60

T

t-SNE
　implementing　175
target　10
temporal feature creation　124
term frequency-inverse document frequency (tf-idf)　243
text preprocessing　112, 113, 116
training and scoring time
　difference between　129
　k-NN, implementing　136
　measure　130, 131
　performance statistics, visualizing　134, 136
　Python script, profiling line by line　137
　Python, code profiling　132
transformers　208

Tree-Based Pipeline Optimization Tool (TPOT)　30
trimming　101
true negative rate (TNR)　55
True Positive (TP)　55
true positive rate (TPR)　56
tuning the hyperparameters　64

U

univariate feature selection　118
unsupervised autoML
　about　152
　agglomerative clustering algorithm, implementing　164
　automation　166, 169
　clustering algorithms, using　153
　components addition, for enhancing pipeline　178
　DBSCAN algorithm, implementing　163
　high-dimensional datasets, visualizing　169, 170
　K-means algorithm, implementing　158, 163
　Principal Component Analysis (PCA), implementing　171, 174
　sample datasets, creating with sklearn　154, 158
　t-SNE, implementing　175
utility function 181

W

weighted links　217
wrapper function
　writing, to optimize XGBoost algorithm　199

X

XGBoost　151

Z

Z- Score standardization　82